ELECTRIC LIGHTING OF THE 20s & 30s

Volume 2

Edited By

James Edward Black

Copyright
1990

Published By
L-W BOOKS
Box 69
Gas City, In 46933

ELECTRIC LIGHTING OF THE 20S & 30S VOL. 1

BY JAMES EDWARD BLACK

132 PAGES COVERING LAMPS OF JEANETTE, HANDEL, MOE BRIDGES, ART DECO, WICKER, FLOOR, TABLE AND NOVELTY LAMPS. 24 LAMPS IN COLOR. ILLUSTRATIONS & PICTURES FROM ORIGINAL CATALOGS.(NO GUESSING IF SHADE MATCHES BASE) PRICES AND ILLUSTRATES 100S OF LAMPS.

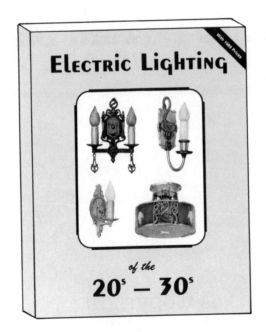

SEND $12.95 PLUS $2.00 PER BOOK SHIPPING TO:
L-W BOOK SALES
P.O. BOX 69
GAS CITY, IN 46933

TABLE OF CONTENTS

Other books available on Lighting

Electric Lighting of the 20s & 30s vol. 1, prices, L-W Publ.	$12.95
Early 20th Century Lighting, Sherwoods Ltd., prices, Schiffer	$16.95
Oil Lamps, Kerosene Era, Thuro, prices	$38.95
Oil Lamps vol. II, KeroseneEra, Thuro, prices	$19.95
Miniature Lamps I, Smith, Schiffer Publ.	$29.95
Miniature Lamps II, Smith, Schiffer Publ.	$28.50
Accurate Price Guide to Min. Lamps I & II, Schiffer Publ.	$10.00
Whale Oil Lamps & Access., Sandwich Glass, Barlow	$24.95
Kerosene Lamps & Access., Sandwich Glass, Barlow-Kaiser	$24.95
Aladdin Electric Lamps, Courter,	$24.95
Aladdin Electric Lamps Price Guide, to above	$5.95
Aladdin: The Magic Name in Lamps, Courter	$15.95
Price Guide to Above	$3.95
19th Century Lighting, Bacot, Candled Lighting	$59.95
Art Nouveau Lamps & Fixtures, Wray	$24.95
American Lighting: 1840-1940, Maril, Domestic Electric	$29.95

FIND THESE BOOKS AT YOUR LOCAL BOOKSTORE OR ORDER FROM:

**L-W BOOK SALES
P. O. BOX 69
GAS CITY, IN 46933
(317) 674-6450**

EMPIRE MADE IN CHICAGO LAMPS

EXPLANATION AND INSTRUCTIONS TO BE OBSERVED

NUMBERS IMPORTANT

In ordering lamps always specify our catalogue number. This will enable us to work with more speed on your order and eliminate many errors.

CONSTRUCTION

Empire lamps are made of a special alloy perfected by us and known as BRITTANIA METAL. The big advantage in our metal is its ability to receive and hold the finishes that we apply. Because of the lasting and wonderful finishes that can be produced on BRITTANIA METAL we consider it the secret to Empire success.

FINISHES

Under each lamp we give a "suggested finish" which is the finish that we recommend for that lamp.

However, you may make your own selection as we can produce any lamp in this catalogue, except where noted, in the following finishes at no additional cost:

Antique Brass
Dull Cyprian Bronze
Dull Peacock
Old Gold
Empire Gold
Verde Antique
Statuary Bronze
Ivory
Polychrome
Brushed Brass and Black
Egyptian Bronze
Olympian Gold
Icynian Bronze

All orders received not specifying any finish will be shipped in the finish that we suggest.

GAS OR ELECTRIC

All the illustrations in this book show our lamps wired for electricity. However, we can furnish any one of them made up for gas without any additional cost, with the exception of those shown on pages 41 to 48, inclusive.

Unless orders distinctly specify "for gas" we will furnish electric in all cases.

FOR GAS

When ordered for gas we will supply a 6 foot hose, burner mantle and chimney, instead of the wiring shown in illustrations at the same prices of electric.

FOR ELECTRIC

Electric lamps always are wired complete with 6 feet silk parallel cord, two pull chain sockets and attachment plug.

RETURNED GOODS

Under no circumstances will we accept goods returned without instructions and permission from us.

BOXING AND CARTAGE

All boxes, casks and barrels are charged at cost. Kindly remember this in placing your order, so as to avoid misunderstanding. We make no charge for cartage.

COLORS OF GLASS

The art glass used is the very best that we can secure. Amber is the color used in all lamps. Amber has, by scientific tests been proven to be easiest on the eyes, and most restful when reading.

TERMS

Our terms are 2% 10 days—30 days net.

CREDIT

Unless you know yourself to be favorably reported by the Commercial Agencies, we suggest that a remittance or satisfactory references accompany your first order. The time incident to the determination of credit standing will be saved by attention to the above request.

C. O. D. SHIPMENTS

We will make C. O. D. express shipment or freight shipments sight draft attached to B/L, only when 25% of amount of invoice is sent with order. When requesting shipment with draft attached to B/L, kindly advise what bank you prefer and we will ask bank to hold draft until arrival of goods.

CLAIMS FOR LOSS OR DAMAGE IN TRANSIT

We take great care in filling orders promptly and packing goods properly, therefore, WE ARE NOT RESPONSIBLE FOR GOODS DAMAGED, OR LOST IN TRANSPORTATION after obtaining receipt from the carriers for delivery to them in good condition. All possible precautions, however, will be used to prevent injury or delay, and shipment will be traced on request. If goods are received in bad order, call this to the attention of the carrier's agent, secure the indorsement of the damage on your bill and file claim with the carrier at once.

CLAIMS FOR SHORTAGE

All our goods are checked by the order clerk and also by the packer, and their counts and weights must tally before shipment is made. All claims for shortage or any other claims must be made in writing three days after receipt of goods. Otherwise we cannot consider your claim.

Empire Lamp & Brass Mfg. Co., 426 S. Clinton St., Chicago, U. S. A.

This beautiful twin shade chandelabra is suitable for office or library

7 inch Shade, 22 inches high

Number E500, Price

Finish suggested, Egyptian Bronze and Polychrome

18 inch Shade, 27 inches high

Number E1070, Price

Finish suggested, Polychrome

18 inch Shade, 27 inches high

Number E470-680, Price

Finish suggested, Old Gold

18 inch Shade, 26 inches high

Number E1100, Price

Finish suggested, Polychrome

18 inch Shade, 26 inches high

Number E1080, Price

Finish suggested, Polychrome

19 inch Shade, 25 inches high

Number E910, Price

Finish suggested, Dull Cyprian Bronze

17 inch Shade, 23 inches high

Number E460, Price

Finish suggested, Egyptian Bronze

18 inch Shade, 26 inches high

Number E470, Price

Finish suggested, Olympian Gold

18 inch Shade, 24 inches high

Number E490, Price

Finish suggested, Dull Cyprian Bronze

18 inch Shade, 26 inches high

Number E610, Price

Finish suggested, Dull Cyprian Bronze

19 inch Shade, 26 inches high

Number E550, Price

Finish suggested, Olympian Gold

18 inch Shade, 24 inches high

Number E940, Price

Finish suggested, Icynian Bronze

18 inch Shade, 22 inches high

Number E950, Price

Finish suggested, Olympian Gold

18 inch Shade, 26 inches high

Number E430, Price

Finish suggested, Dull Cyprian Bronze

18 inch Shade, 25 inches high

Number E440, Price

Finish suggested, Olympian Gold

19 inch Shade, 25 inches high

Number E770, Price

Finish suggested, Antique Brass

18 inch Shade, 25 inches high

Number E820, Price

Finish suggested, Dull Peacock

18 inch Shade, 27 inches high
Number E960, Price
Finish suggested, Olympian Gold

18 inch Shade, 24 inches high
Number E1060, Price
Finish suggested, Polychrome

16 inch Shade, 22 inches high
Number E570, Price
Finish suggested, Verde Antique

16 inch Shade, 22 inches high
Number E620, Price
Finish suggested, Dull Cyprian Bronze

18 inch Shade, 25 inches high
Number E322, Price
Finish suggested, Old Gold

20 inch Shade, 27 inches high
Number E450, Price
Finish suggested, Olympian Gold

Empire Quality Lamps
Made in Chicago

20 inch Shade, 26 inches high
Number E850, Price
Finish suggested, Statuary Bronze

18 inch Shade, 23 inches high
Number E880, Price
Finish suggested, Polychrome

Empire Quality Lamps
Made in Chicago

18 inch Shade, 23 inches high

Number E800, Price

Finish suggested, Dull Cyprian Bronze

18 inch Shade, 23 inches high

Number E120, Price

Finish suggested, Icynian Bronze

21 inch Shade, 26 inches high
Number E540, Price
Finish suggested, Olympian Gold

18 inch Shade, 25 inches high
Number E580, Price
Finish suggested, Olympian Gold

18 inch Shade, 22 inches high

Number E810, Price

Finish suggested, Olympian Gold

18 inch Shade, 24 inches high

Number E760, Price

Finish suggested, Dull Cyprian Bronze

18 inch Shade, 24 inches high

Number E590, Price

Finish suggested, Verde Antique

18 inch Shade, 23 inches high

Number E600, Price

Finish suggested, Olympian Bronze

16 inch Shade, 23 inches high
Number E560, Price
Finish suggested, Brush Brass and Black

18 inch Shade, 23 inches high
Number E930. Price
Finish suggested, Icynian Bronze

17 inch Shade, 25 inches high
Number E1040, Price
Finish suggested, Dull Cyprian Bronze and
Polychrome

18 inch Shade, 23 inches high
Number E1090, Price
Finish suggested, Polychrome

10

Empire Quality Lamps
Made in Chicago

18 inch Shade, 23 inches high

Number E1020, Price

Finish suggested, Dull Peacock

18 inch Shade, 24 inches high

Number E1030, Price

Finish Suggested, Egyptian Bronze

17 inch Shade, 25 inches high

Number E900, Price

Finish suggested, Dull Cyprian Bronze

17 inch Shade, 24 inches high

Number E790, Price

Finish suggested, Brush Brass and Black

18 inch Shade, 24 inches high

Number E630, Price

Finish suggested, Old Gold

18 inch Shade, 23 inches high

Number E710, Price

Finish suggested, Dull Cyprian Bronze

18 inch Shade, 27 inches high

Number E680, Price

Finish suggested, Olympian Gold

19 inch Shade, 24 inches high

Number E970, Price

Finish suggested, Polychrome

16 inch Shade, 23 inches high

Number E140, Price

Finish suggested, Antique Brass

18 inch Shade, 24 inches high

Number E85, Price

Finish suggested, Old Gold

18 inch Shade, 28 inches high

Number E170, Price

Finish suggested, Antique Brass

18 inch Sq. Shade, 25 inches high

Number E50, Price

Finish suggested, Olympian Gold

18 inch Shade, 25 inches high

Number E70, Price

Finish suggested, Olympian Gold

18 inch Hexagon Shade, 24 inches high

Number E80, Price

Finish suggested, Brush Brass and Black

17 inch Shade, 23 inches high

Number E90, Price

Finish suggested, Olympian Gold

18 inch Shade, 23 inches high

Number E100, Price

Finish suggested, Polychrome

20 inch Shade, 25 inches high

Number E230, Price

Finish suggested, Verde Antique

18 inch Shade, 28 inches high

Number E250, Price

Finish suggested, Polychrome

18 inch Shade, 23 inches high

Number E210, Price

Finish suggested, Olympian Gold

17 inch Shade, 23 inches high

Number E220, Price

Finish suggested, Statuary Bronze

18 inch Shade, 24 inches high
Number E180, Price
Finish suggested, Olympian Gold

18 inch Shade, 26 inches high
Number E60, Price
Finish suggested, Brush Brass and Black

16 inch Shade, 22 inches high
Number E321, Price
Finish suggested, Icynian Bronze

18 inch Shade, 25 inches high
Number E280, Price
Finish suggested, Polychrome

16

16 inch Shade, 22 inches high

Number E150, Price

Finish suggested, Dull Cyprian Bronze

20 inch Shade, 23 inches high

Number E130, Price

Finish suggested, Verde Antique

19 inch Shade, 24 inches high
Number E190, Price
Finish suggested, Statuary Bronze

18 inch Shade, 24 inches high
Number E200, Price
Finish suggested, Brush Brass and Black

Empire Quality Lamps
Made in Chicago

16 inch Shade, 22 inches high
Number E240, Price
Finish suggested, Olympian Gold

17 inch Shade, 24 inches high
Number E270, Price
Finish suggested, Icynian Bronze

16 inch Sq. Shade, 24 inches high
Number E40, Price
Finish suggested, Egyptian Bronze and
Polychrome

20 inch Shade, 24 inches high
Number E30, Price
Finish suggested, Olympian Gold

18

18 inch Shade, 22 inches high

Number E810-90, Price

Finish suggested, Polychrome

18 inch Shade, 23 inches high
Number E160, Price
Finish suggested, Brush Brass and Black

18 inch Shade, 23 inches high

Number E260, Price

Finish suggested, Old Gold

17 inch Shade, 22 inches high

Number E110, Price

Finish suggested, Dull Cyprian Bronze

17 inch Shade, 22 inches high

Number E290, Price

Finish suggested, Statuary Bronze

18 inch Shade, 25 inches high

Number E1050, Price

Finish suggested, Polychrome

19 inch Shade, 24 inches high
Number E750, Price
Finish suggested, Icynian Bronze

17 inch Shade, 22 inches high
Number E780, Price
Finish suggested, Dull Cyprian

17 inch Shade, 22 inches high
Number E1010, Price
Finish suggested, Dull Cyprian Bronze

18 inch Shade, 27 inches high
Number E901, Price
Finish suggested, Egyptian Bronze

17 inch Shade, 24 inches high

Number E20-100, Price

Finish suggested, Antique Brass

12 inch Shade, 18 inches high
Number E5, Price
Finish suggested, Ivory

17 inch Shade, 22 inches high

Number E10, Price

Finish suggested, Dull Cyprian Bronze

17 inch Shade, 24 inches high

Number E20, Price

Finish suggested, Dull Cyprian Bronze

14 inch Shade, 17 inches high
Number E380, Price
Finish suggested, Brush Brass and Black

18 inches high
Number E345B, Price

15 inches high
Number E370B, Price

These Bases can be used with most any
2¼ inch silk or glass shade. Specify
in ordering whether Ivory, Old Gold
or Verde Antique finish is desired.

Here's the famous Victory Lamp. Double adjustments and positions. The
novelty light with the Doughboy hat
Number 3750, Price
Finish, Brush Brass

The "Utilite." This little lamp can be
used in a dozen ways, and is a big seller.
Number 3724, Price
Finish, Brush Brass

14 inches high
Number E335B, Price

15 inches high
Number E340B, Price

The famous "Portalite" is the biggest selling novelty lamp of the season. The
lowest priced quality adjustable lamp ever produced.
Number, 11021, Price   Finish, Brush Brass

21

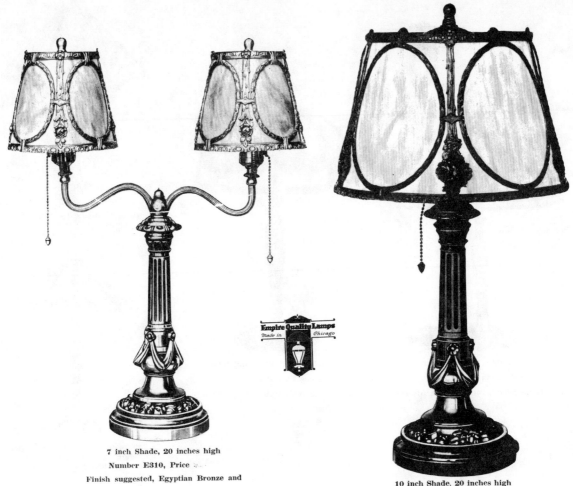

7 inch Shade, 20 inches high

Number E310, Price

Finish suggested, Egyptian Bronze and
Polychrome

10 inch Shade, 20 inches high
Number E320, Price
Finish suggested, Egyptian Bronze and Polychrome

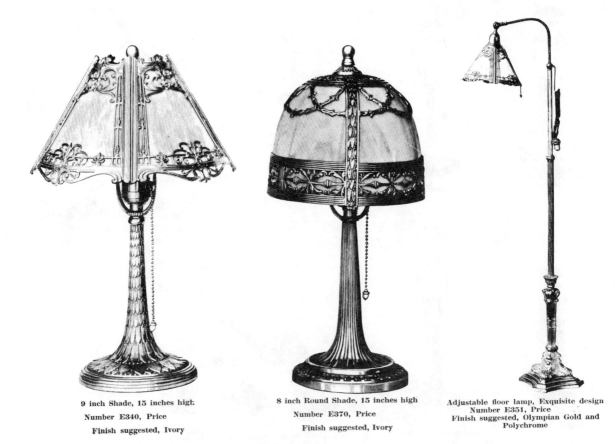

9 inch Shade, 15 inches high

Number E340, Price

Finish suggested, Ivory

8 inch Round Shade, 15 inches high

Number E370, Price

Finish suggested, Ivory

Adjustable floor lamp, Exquisite design
Number E351, Price
Finish suggested, Olympian Gold and
Polychrome

6 inch Square Shade, 15 inches high

Number E410, Price

Finish suggested, Verde Antique

6 inch Round Shade, 18 inches high
Number E330, Price
Finish suggested, Olympian Gold and
Polychrome

Adjustable Table or Desk Lamp

8 inch Oblong Shade, 17 inches high

Number E390, Price

Finish suggested, Old Gold

Twin Adjustment Piano Lamp
9 inch Shade
Number E402, Price
Finish suggested, Old Gold

Twin adjustment piano and desk lamp
8 inch Shade
Number E401, Price
Finish suggested, Old Gold

Twin Adjustment Piano or Desk Lamp
6 inch Shade
Number E420, Price
Finish suggested, Icynian Bronze

Twin adjustment piano and table lamp
8 inch Shade
Number E400, Price
Finish suggested, Polychrome

Universal adjustment floor reading lamp
10 inch Shade, extends 72 inches in over all
Number E350, Price
Finishes suggested, Antique Brass, Statuary Bronze, Olympian Gold, Verde Antique

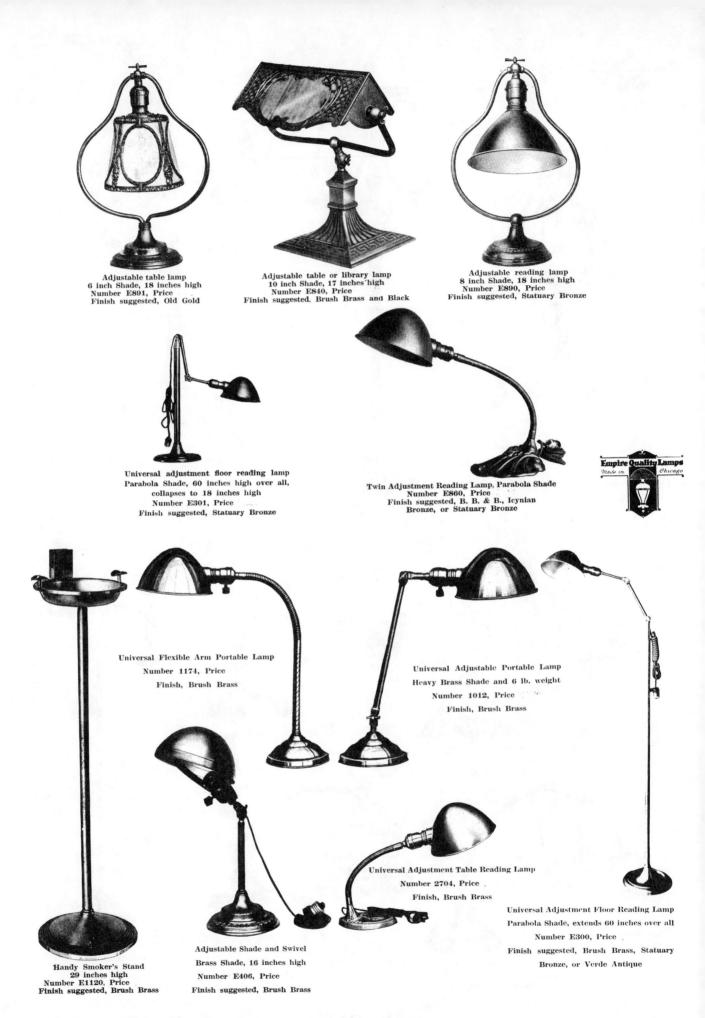

Adjustable table lamp
6 inch Shade, 18 inches high
Number E891, Price
Finish suggested, Old Gold

Adjustable table or library lamp
10 inch Shade, 17 inches high
Number E840, Price
Finish suggested, Brush Brass and Black

Adjustable reading lamp
8 inch Shade, 18 inches high
Number E890, Price
Finish suggested, Statuary Bronze

Universal adjustment floor reading lamp
Parabola Shade, 60 inches high over all,
collapses to 18 inches high
Number E301, Price
Finish suggested, Statuary Bronze

Twin Adjustment Reading Lamp, Parabola Shade
Number E860, Price
Finish suggested, B. B. & B., Icynian
Bronze, or Statuary Bronze

Empire Quality Lamps
Made in Chicago

Universal Flexible Arm Portable Lamp
Number 1174, Price
Finish, Brush Brass

Universal Adjustable Portable Lamp
Heavy Brass Shade and 6 lb. weight
Number 1012, Price
Finish, Brush Brass

Handy Smoker's Stand
29 inches high
Number E1120, Price
Finish suggested, Brush Brass

Adjustable Shade and Swivel
Brass Shade, 16 inches high
Number E406, Price
Finish suggested, Brush Brass

Universal Adjustment Table Reading Lamp
Number 2704, Price
Finish, Brush Brass

Universal Adjustment Floor Reading Lamp
Parabola Shade, extends 60 inches over all
Number E300, Price
Finish suggested, Brush Brass, Statuary
Bronze, or Verde Antique

24

NUMBER S6-162

Design patent

Length, 8 inches.   4½x8-inch oval canopy

Socket covers for keyless sockets included

24 hour shipments

Antique brush brass finish

Not Wired   Complete

Electric ....

Shade No. B-1110 white satin finish

NUMBER S6-164

Design patent

Length, 12 inches. Diameter of plate, 8½ inches

Socket covers for keyless sockets included

24 hour shipments

Antique brush brass finish

Not Wired   Complete

2 Electric ....

Shade No. B-145 white satin finish

NUMBER S6-163

Design patent

Length, 9 inches.   Diameter of plate, 6 inches

24 hour shipments

Antique brush brass finish

Not Wired   Complete

2 Electric ....

Shade No. B-1110 white satin finish

THIS MARK ON
B
TRADE MARK
REGISTERED
U.S. PAT. OFF.
ALL CHANDELIERS

NUMBER S6-165

Design patent

Length, 24 inches. Oval plate, 7¼x13¾ inches

4-inch cut glass prisms included

24 hour shipments

Antique brush brass finish

Not Wired   Complete

1 Electric ....

Pull sockets included in complete price

NUMBER S6-166

Design patent

Length, 24 inches. Oval plate, 7¼x13¾ inches

Socket covers for key sockets included

24 hour shipments

Antique brush brass finish

Not Wired   Complete

2 Electric ....

Shade No. B-145 white satin finish

THIS MARK ON
B
TRADE MARK
REGISTERED
U.S. PAT. OFF.
ALL CHANDELIERS

NUMBER S6-212
For 60-watt lamp
Length, 18 inches
No. 1202, 7¼-inch shade and holder included
24 hour shipments
Brush brass finish
Not Wired Complete
1 Electric . . .

NUMBER S6-213
For 100-watt lamp
No. 1203, 8-inch shade
Not Wired Complete
1 Electric . . .

NUMBER S6-214
For 60-watt lamp
Length, 18 inches
Six-inch bent alabaster shade and holder included
24 hour shipments
Brush brass finish
Not Wired Complete
1 Electric . . .

NUMBER S6-215
For 60-watt lamp
Design patent
Length, 18 inches
No. B-11710, 8-inch white etched shade and holder included
24 hour shipments
Antique brush brass finish
Not Wired Complete
1 Electric . . .

NUMBER S6-216
For 60-watt lamp
Length, 18 inches
No. B-903, 8-inch opal sh
and holder included
24 hour shipments
Brush brass finish
Not Wired Com

NUMBER S6-217
For 100-watt lamp
No. B-904, 10-inch opal sh
Not Wired Com
1 Electric . . .

NUMBER S6-218
Length, 16 inches
24 hour shipments
Brush brass finish

NUMBER S6-219
Length, 16 inches
24 hour shipments
Brush brass finish

NUMBER S6-220
Length, 16 inches
24 hour shipments
Brush brass finish

NUMBER S6-221
Design patent
Length, 15 inches
24 hour shipments
Antique brush brass fi

NUMBER S6-323

Extends, 4 inches

4-inch cut glass prisms included

24 hour shipments

Antique brush brass finish

Not Wired Complete

1 Electric ....

Complete price includes pull socket

NUMBER S6-332

Design patent

Extends, 9 inches

24 hour shipments

Antique brush brass finish

Not Wired Complete

1 Electric ....

Shade No. B-145 white satin finish

NUMBER S6-333

Extends, 7 inches

Socket cover for key socket included

24 hour shipments

Brush brass finish

Not Wired Complete

1 Electric ....

Shade No. B-1804 white satin finish

NUMBER S6-326

Extends, 4 inches

Socket cover for key socket included

24 hour shipments

Brush brass finish

Not Wired Complete

1 Electric ....

Shade No. B-1058 white satin finish

THIS MARK ON
B
TRADE MARK
REGISTERED
U.S. PAT. OFF.
ALL CHANDELIERS

NUMBER S6-335

Design patent

Extends, 6 inches

24 hour shipments

Antique brush brass finish

Not Wired Complete

1 Electric ....

Shade No. B-145 white satin finish

NUMBER S6-336

Extends, 7 inches

Socket husk for key socket included

Lamp not furnished

24 hour shipments

Brush brass finish

Not Wired Complete

1 Electric ....

NUMBER S6-329

Extends, 4 inches

Socket cover for key socket included

24 hour shipments

Brush brass and black finish

NUMBER S6-338

Extends, 9½ inches

24 hour shipments

Brush brass finish

Not Wired Complete

1 Electric ....

Shade B-500 crystal roughed inside

NUMBER S6-339

Design patent

Extends, 6 inches

24 hour shipments

Antique brush brass finish

27

THIS MARK ON
B
TRADE MARK
REGISTERED
U.S. PAT. OFF.
ALL CHANDELIERS

NUMBER S6-389

7-inch frosted aluminum shade included

24 hour shipments

Brush brass finish

Not Wired Complete

1 Electric ....

Attachment plug included in complete price

NUMBER S6-390

Height, 17 inches

24 hour shipments

Brush brass finish

Complete with green glass shade, wire, pull chain socket and attachment plug.

NUMBER S6-391

7-inch frosted aluminum sha included

24 hour shipments

Brush brass finish

Not Wired Co

1 Electric ....

Attachment plug included complete price

NUMBER S6-392

24 hour shipments

Brush brass finish

Complete with green glass shade, wire, pull chain socket and attachment plug.

NUMBER S6-394

24 hour shipments

Brush brass finish

Complete with sheet metal shade, wire, pull chain socket and attachment plug.

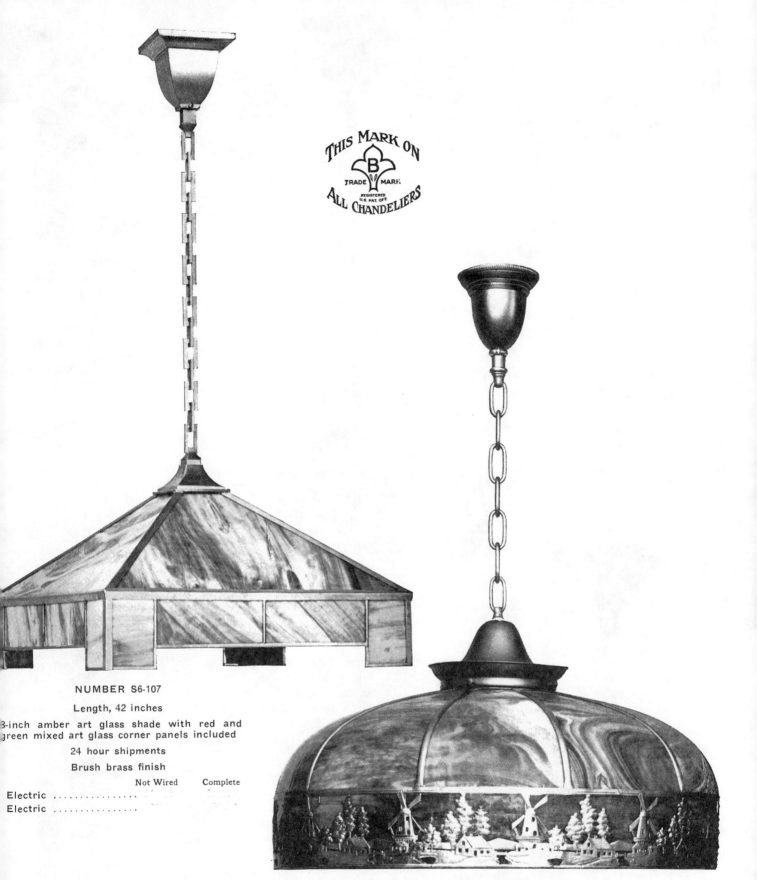

NUMBER S6-107

Length, 42 inches

3-inch amber art glass shade with red and
green mixed art glass corner panels included

24 hour shipments

Brush brass finish

	Not Wired	Complete
Electric		
Electric		

NUMBER S6-108

Length, 42 inches

24-inch bent amber art glass shade, sunset
color apron panels with metal
overlay included

24 hour shipments

Brush brass finish

NUMBER S6-375

Extends, 6 inches

Lantern with amber art glass
panels included

24 hour shipments

Brush brass finish

Not Wired Complete

1 Electric ....

Complete price includes pull chain
socket

THIS MARK ON
B TRADE MARK
REGISTERED
U.S. PAT. OFF.
ALL CHANDELIERS

NUMBER S6-376

Extends, 6 inches

Lantern with C. R. I. glass inclu

24 hour shipments

Brush brass finish

Not Wired Comp

1 Electric ....

Complete price includes pull ch
socket

NUMBER S6-377

Length, 36 inches. Spread, 15 inches

24 hour shipments

Brush brass finish

Not Wired Complete

2 Electric ....
4 Electric ....

Shade No. B-9547 C. R. I

NUMBER S6-378

Length, 16 inches

Lantern with amber art glass panels
included

24 hour shipments

Brush brass finish

NUMBER S6-379

Length, 15 inches

Lantern with C. R. I. glass inclu

24 hour shipments

Brush brass finish

NUMBER S6-131

Length, 30 inches

t glass shade with green
uter leaves and pink in-
side leaves, and socket
cover for key socket
included

24 hour shipments

Brush brass finish

Not Wired Complete

Electric . . .

NUMBER S6-132

Length, 30 inches

Amber art glass lantern, sunset color lower panel with
metal overlay, and socket
cover for keyless socket
included

NUMBER S6-133

Length, 30 inches

Amber art glass lantern with
cathedral art glass design,
and socket cover for key
socket included

24 hour shipments

Brush brass finish

Not Wired Complete

1 Electric . . .

NUMBER S6-134

Length, 30 inches

Alabaster art glass lantern
with amber art glass de-
sign and socket cover
for keyless socket
included

31

THIS MARK ON
ALL CHANDELIERS
TRADE MARK
B
REGISTERED
U.S. PAT. OFF.

NUMBER S6-109

Length, 42 inches

24-inch amber art glass shade, sunset color
apron panels with metal overlay included

24 hour shipments

Brush brass finish

NUMBER S6-110

Length, 42 inches

22-inch amber art glass shade, with blended
red and green border panels, ruby corners
and diamonds, and pink design in
center of apron panel included

32

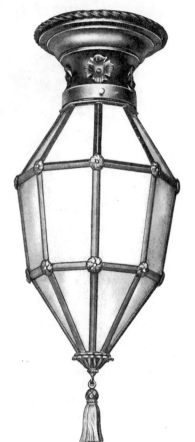

Design A-107

Diameter of Ceiling Ring,
7½ inches.

Length, 22 inches (over all).

Includes Lantern, 4x14 inches.

With White Cathedral Glass,
Satin Finish.

Rusty Iron Polychrome.

Electric, complete as
shown ....

Design A-106

Diameter of Ceiling Ring,
7 inches.

Length, 12 inches (over all).

Includes G-459, 3¼x9½-inch
Lantern.

With White Cathedral Glass,
Satin Finish.

Rusty Iron Polychrome.

Electric, complete as
shown ....

Design A-108

Length, 36 inches.

Includes Lantern, 5 x 12 inches.

With White Cathedral Glass, Satin Finish.

Rusty Iron Polychrome.

Electric, complete as shown.....

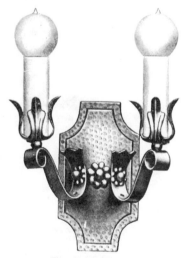

Design A-110

Extends, 4 inches

Spread, 6½ inches.

Cast Back Plate, 4¼x7½ inches.

Includes Canopy Switch.

Lamps not furnished.

Design A-109

Extends, 3 inches.

Cast Back Plate, 4½x9 inches.

Includes Chain Pull Socket.

Lamp not furnished.

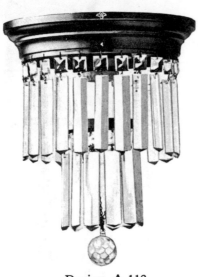

Design A-112

Diameter of Spinning, 10 inches.

Length, 10 inches (over all).

Includes 4-inch Colonial Cut Prisms and
Cut Glass Bottom Ball.

Antique Brush Silver.

1 Electric, complete as shown...

THIS MARK ON
ALL CHANDELIERS

B TRADE MARK
REGISTERED
U.S. PAT. OFF.

Design A-118

Length, 36 inches.

Includes Clear Glass Cylinder,
8x10 inches.

Lamps not furnished.

Colonial Brass.

3 Electric, complete as shown....

Extra lengthening, per foot,

Design A-115

Design A-113

Length, 36 inches.

Includes 12-inch Florentine Etche
and Old Gold Silk Tassel.

Colonial Gold, Brown Relief.

1 Electric, complete as shown....

Extra lengthening, per foot (3 c

THIS MARK ON
ALL CHANDELIERS
B TRADE MARK
REGISTERED U.S. PAT. OFF.

Another example of the semi-indirect and candle illumination. The graceful supporting arms of reeded tube terminate in a single chain hanger. A pleasing design suitable for the living or dining room. Bowl is etched calcite glass. The wall brackets are designed to match. Shown on plate 421.

Design A-51

Length, 36 inches. Includes 16-inch Calcite Bowl.

Colonial Gold, Brown Relief.

Electric, complete as shown...............

Extra lengthening, per foot,

Semi-indirect illumination retains its popularity when consistently designed. The ones illustrated are exceptionally attractive. The manufacture of calcite glass marks an epoch in the art of the glass maker. Built up in layers, the inner layer or reflecting surface is of polished opal glass, the outer of a pearl iridescent hue. Between the two is a layer of brown, imparting to the bowl, when lighted, a rich golden tone.

Length, 42 inches. Spread, 21 inches.

Cast Oval Arms.

14-inch Calcite Bowl and Old Gold Color Silk Shades included.

Lamps not furnished.

Antique Brush Silver.

Design A-49

Design A-86

Extends 12 inches.

Includes Old Rose Silk Shade, 10 inches in diameter, and Chain Pull Socket.

Colonial Brass.

1 Electric, complete as shown...

Length, 7 inches (over all).
Diameter of Plate, 11 inches.
Lamps not furnished.
Antique Ivory.

Design A-92

3 Electric, complete as shown.....

Design A-93

4 Electric, complete as shown...

Design A-138

Diameter of Canopy, 4½ inches.
Lamp not furnished.

Polished Nickel.

1 Electric, complete as shown.....

Design A-88

Length, 36 inches.

Includes Old Rose Silk Shade, 10 inches in diameter, and Chain Pull Socket.

Colonial Brass.

Design A-59

Ceiling Ring, 6 inches diameter
Lamp not furnished.

Antique Sand Blast High Lighte

1 Electric, complete as shown.....

THIS MARK ON ALL CHANDELIERS

TRADE MARK
REGISTERED
U.S. PAT. OFF.

NUMBER S6-17

Length, 36 inches.   Spread, 14½ inches

Socket covers for key sockets included

24 hour shipments

Brush brass finish

                    Not Wired   Complete

3 Electric ....

4 Electric ....

Shade B-124 white satin finish

Crystal trimmed chandeliers have always been popular for the more formal rooms. Many attractive designs are incorporated in this catalog.

NUMBER S6-18

Design patent

Length, 36 inches.   Spread, 16 inches

Colonial cut prisms included

Cast brass arms

24 hour shipments

Antique brush brass finish

Suggestions
for the
Bedroom, Hall, Bathroom and Kitchen

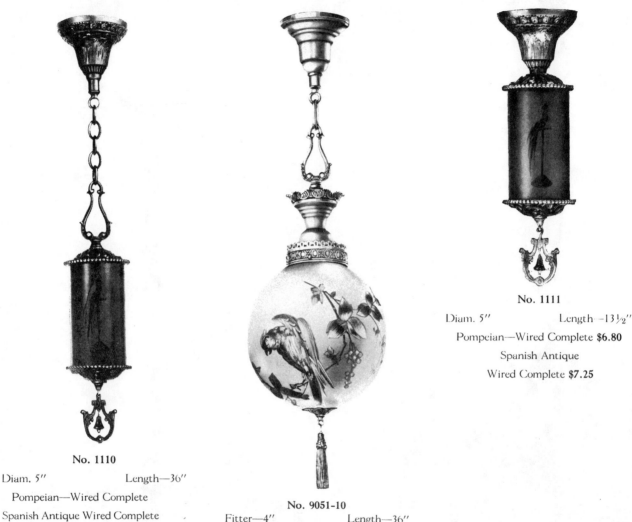

No. 1110

Diam. 5″   Length—36″

Pompeian—Wired Complete

Spanish Antique Wired Complete

No. 9051-10

Fitter—4″   Length—36″

Fixture only—Bare   Wired

Decorated in Pompeian

An unusual piece
for
Vestibules, Halls and Sunrooms

9051-10 with glassware as shown Bare   Wired

No. 1111

Diam. 5″   Length—13½″

Pompeian—Wired Complete **$6.80**

Spanish Antique

Wired Complete **$7.25**

Fixtures in this folder can be furnished only in the finishes shown

Made By

Diamond "F" Manufacturing Co.
2614-22 East 51st Street
Cleveland, Ohio

No. 640
Height—65″
Finish: Swedish Iron
Code: *Wade*

No. 525
Height—65″
Finish: Antique Gold
Code: *Wager*

No. 650
Height—67″
Finish: Swedish Iron
Code: *Wail*

No. 630
Height—65″
Finish: Swedish Iron
Code: *Walk*

No. 545
Height—66″
Finish:
Plated Gold Lacquered
Code: *Wall*

No. 2602

Spread—8″   Back—10x5½″
Complete with Flametint
Candalite Bulb
Code: Cruise

No. 2601

Extends—5″   Back—10x5½″
Complete with Flametint
Candalite Bulb
Code: Crust

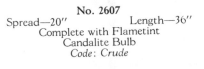

No. 2607

Spread—20″   Length—36″
Complete with Flametint
Candalite Bulb
Code: Crude

PLEASINGLY NEW AND
DIFFERENT

(PATENT APPLIED FOR)

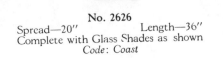

No. 2626

Spread—20″   Length—36″
Complete with Glass Shades as shown
Code: Coast

No. 2606

Spread—20″   Length—36″
Code: Coach

No. 1010-B14—Copper
Flemish Copper Finish
Cathedral Amber Glass
Height 13½″   Extension 7⅛″
Code: *Root*

No. 11004
Open Bottom Copper
Code: *Rude*
No. 11005
Closed Bottom Copper
Code: *Ruin*
Flemish Copper Finish
Cathedral Amber Glass
Length overall—11½″
Width—5½″

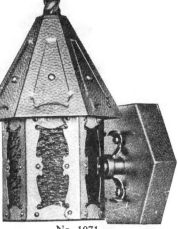

No. 1071
Made of Copper
Extension—6⅝″
Length overall 10″
Lantern 7¾″ long, 5¼″ wide
Bronze Finish
Amber Cathedral Glass
Code: *Rotund*

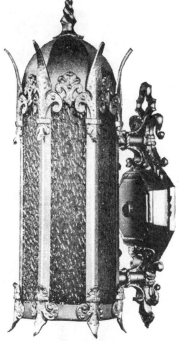

No. 1110-B-31
Made of Copper
Extension—7½″
Length overall—17½″
Lantern 17½″ long, 5¾″ wide
Bronze Finish
Amber Cathedral Glass
Code: *Roman*

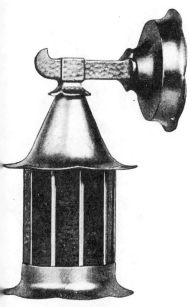

No. 11001
Open Bottom Copper
Code: *Rub*
No. 11002
Closed Bottom Copper
Code: *Rubric*
Flemish Copper Finish
Cathedral Amber Glass
Height 11″   Extension 7′

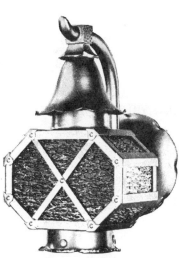

No. 1092-B-29
Made of Copper
Extension—7½″
Length overall—10¾″
Lantern: 9″ long
6½″ wide, 3½″ deep
Bronze Finish
Amber Cathedral Glass
Code: *Rocket*

No. 1120-B-33—Made of Copper
Extension 8″   Length overall 18½″
Lantern 13″ long, 7¼″ wide
Bronze Finish
Amber Cathedral Glass   Code: *Royal*

No. L-1074
Made of Copper
Length overall 11″
Lantern 7¾″ long, 5¼″ wid
Bronze Finish
Amber Cathedral Glass
Code: *Robust*

No. 1905—5 Lights
Spread—20″   Length—11″
Code: Garb

No. 1900
Spread—7″   Depth—3″
Code: Gain

No. 1914/66
Spread—7″   Fitter—4″
Complete with Glass
Decorated Pongee
9½″ x 7″ x 4″
Code: Gaze
No. 1914
Holder only, less glass.
Code: Gay

MELROSE SERIES

Fixtures in this series are made in two finishes:

MELROSE BRASS as shown on the color plate on the next page, and MELROSE GOLD—Antique Gold is the basic color, artistically relieved with tints of Nile Green and Carmine red, properly blended together, producing a richly subdued combination of colors. Warm in appearance and pleasing to the eye.

These finishes will retain their beauty indefinitely.

ATTRACTIVE IN DESIGN
AND FINISH

No. 1915/909
Length—36″   Fitter—4″
Complete with Basket Bowl
Decorated to Match
10″ x 4″
Code: Giant
No. 1915
Hanger Only—Less Glass
Code: Gem

No. 1907—5 Lights
Spread—18″   Length—36″
Code: Gaudy

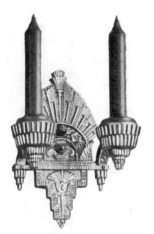

No. 2602
Spread—8″   Back—10x5½″
Complete with Flametint
Candalite Bulb
Code: Cruise

No. 2601
Extends—5″   Back—10x5½″
Complete with Flametint
Candalite Bulb
Code: Crust

No. 2607
Spread—20″   Length—36″
Complete with Flametint
Candalite Bulb
Code: Crude

PLEASINGLY NEW AND
DIFFERENT

(PATENT APPLIED FOR)

No. 2626
Spread—20″   Length—36″
Complete with Glass Shades as shown

No. 2606
Spread—20″   Length—36″

43

Torchiers and candle sticks shown on this page are furnished complete with cord and plug.

No. 95—Dec. 4
Height—12″
Alcazar Finish
Code: Abhor

No. 95—Dec. 5
Height—12″
Alcazar Finish
Code: Abide

No. 95—Dec. 6
Height—12″
Alcazar Finish
Code: Abode

No. 57 Candlesticks
Height overall—15″
Finish—Antique Gold
or Antique Silver
Code: Wander

No. 50 Candlesticks
Height overall—18″
Finish: Antique Gold
or Antique Silver
Code: Ward

No. 118
Torchier
Height—13″
Antique Gold, Amber
Crackled Cylinder
Code: Walnut

No. 66
Candlestick
Decorated in
Pompeian
Code: Weird

No. 150
Torchier
Height—16″
Finish: Gold and Polly
Code: Wear

No. 140
Parrot Lamp
Plated Finish—Crystal Onyx Ball
Height—13″
Code: Warm

No. 153
Newell Post Lamp
Height—18″
Bronze and Green Finish
Code: Weld

No. 160—All Metal Parrot Lamp
Height—13″
Code: Wave

No. 417
Roof Diameter—9¼"   Depth—6½"
Hinged Glass Bottom Door
Material: Copper   Glass: Amber Moss
Code: *Rib*

No. 224
Height overall—19½"   Plate—4½" x 11"
Hinged Bottom Door   Material: Iron
Finish: Black   Glass: Clear Moss
Code: *Rave*

No. U780
Extends—15"   Length overall—28"
Roof Diameter—10½"   Height—20"
Material: Iron
Finish: Black   Glass: Clear Moss
Code: *Rondo*

No. 2518
Roof Diameter—9½"   Depth—6½"
Material: Copper   Interchangeable Numerals
Code: *Rove*

No. BRK2516
Height overall—7¼"   Body—9½" x 4¾"
Material: Copper   Interchangeable Numerals
Code: *Rout*

No. E654
Length overall—33"
Roof Diameter—9"   Height—14½"
Material: Iron
Finish: Black   Glass: Clear Moss
Code: *Rod*

No. BRJCB460
Extends—8½"   Length overall—15"
Roof Diameter—7"   Height—10"
Material: Copper   Glass: Amber Moss
Code: *Rife*

No. 2502
Roof—9½" x 5"   Depth—9¾"
Hinged Bottom Door
Material: Copper   Interchangeable Numerals
Code: *Rouge*

No. AWL454
Length overall—16½"
Roof Diameter—7"   Open Bottom
Material: Copper   Glass: Amber Moss
Code: *Ridge*

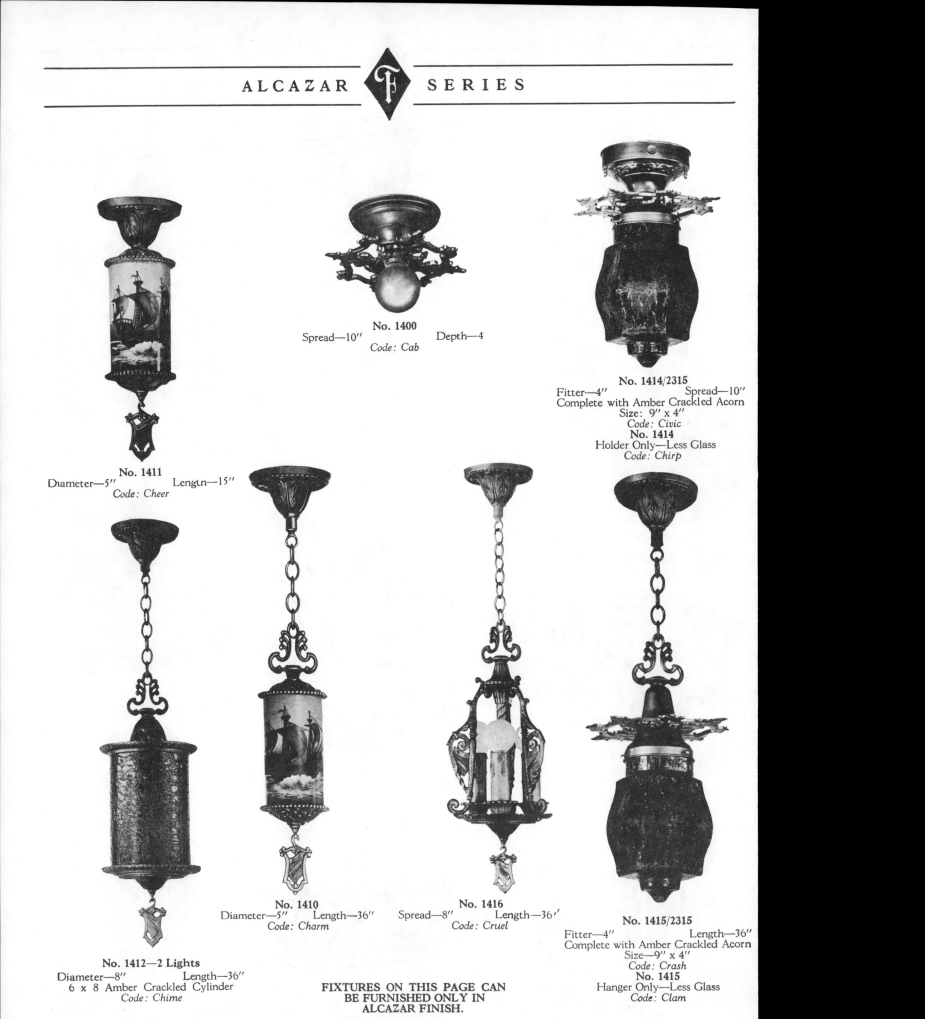

No. 1400

Spread—10"   Depth—4

Code: Cab

No. 1414/2315
Fitter—4"   Spread—10"
Complete with Amber Crackled Acorn
Size: 9" x 4"
Code: Civic
No. 1414
Holder Only—Less Glass
Code: Chirp

No. 1411
Diameter—5"   Length—15"
Code: Cheer

No. 1410
Diameter—5"   Length—36"
Code: Charm

No. 1416
Spread—8"   Length—36"
Code: Cruel

No. 1415/2315
Fitter—4"   Length—36"
Complete with Amber Crackled Acorn
Size—9" x 4"
Code: Crash
No. 1415
Hanger Only—Less Glass
Code: Clam

No. 1412—2 Lights
Diameter—8"   Length—36"
6 x 8 Amber Crackled Cylinder
Code: Chime

**FIXTURES ON THIS PAGE CAN
BE FURNISHED ONLY IN
ALCAZAR FINISH.**

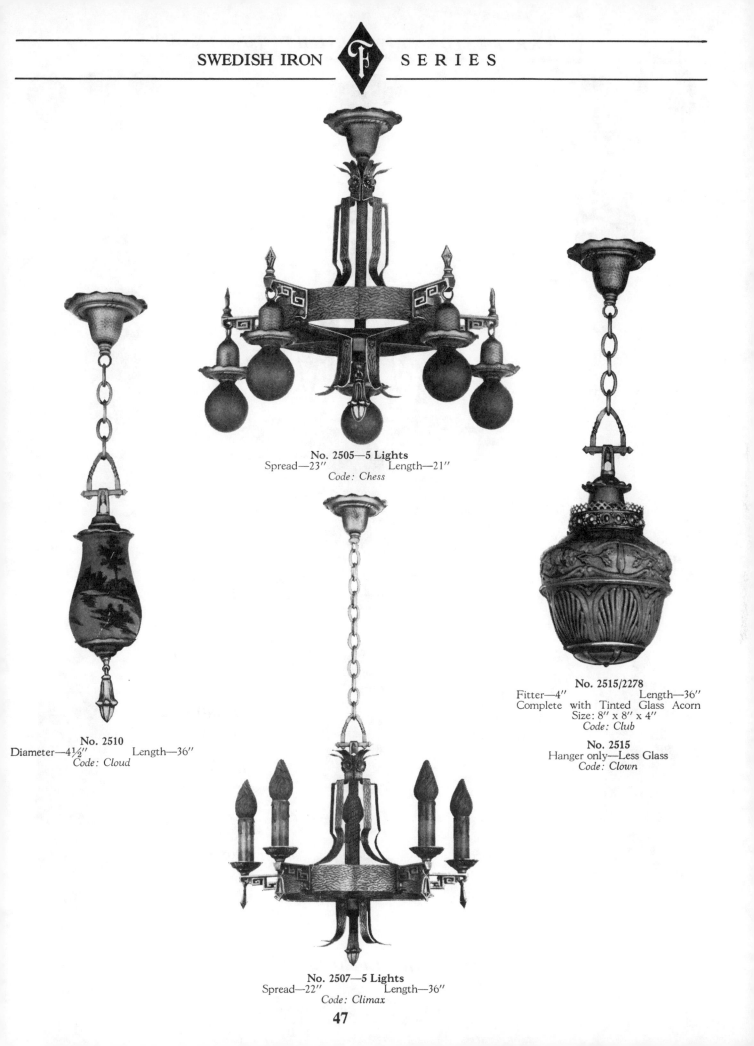

No. 2505—5 Lights
Spread—23″   Length—21″
Code: Chess

No. 2510
Diameter—4½″   Length—36″
Code: Cloud

No. 2515/2278
Fitter—4″   Length—36″
Complete with Tinted Glass Acorn
Size: 8″ x 8″ x 4″
Code: Club

No. 2515
Hanger only—Less Glass
Code: Clown

No. 2507—5 Lights
Spread—22″   Length—36″
Code: Climax

No. 1411

Diameter—5"   Length—15"

Wired complete with Glass **$7.30**

Bare with Glass **$6.70**

FIXTURES ON THIS PAGE CAN BE FURNISHED ONLY IN ALCAZAR FINISH.

No. 1406—5 Lights

Spread—20"   Length—36"

Wired **$24.50**

No. 1410

Diameter—5"   Length—36"

Wired Complete with Glass **$8.50**

Bare with Glass **$7.75**

No. 95—Dec. 4

Height—12"

Each **$7.50**

DECORATIVE DISTINCTIVE DIFFERENT

❦

TORCHIERS FURNISHED COMPLETE WITH SWITCH AND CORD

No. 95—Dec. 6

Height—12"

Each **$7.50**

RICH MASSIVE ELEGANT

❦

TORCHIERS FURNISHED COMPLETE WITH SWITCH AND CORD

No. 95—Dec. 5

Height—12"

Each **$7.50**

Decorative Lighting Fixtures

No. 9078

Extends—5¼"   Height—10"

No. 9079

Spread—8"   Diam.—4½"

TAN GOLD—A beautiful velvet tan blending off into a rich dark mahogany. Shaded with dull gold and artistically trimmed with pleasing colors.

No. 9044—5 Lights

Spread—22"   Length—22"

RUSTIC IRON—A dull black background combined with red and gold coloring. A suggestive brown tinge to the black background produces a warm and colorful rustic effect.

No. 9031

Wired with Sign Receptacle

Spread—10"   Length—12½"

Fixtures on this page Decorated only in Tan Gold and Rustic Iron

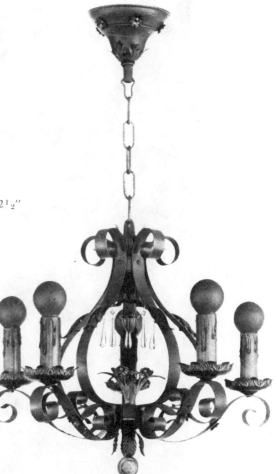

No. 9046—5 Lights

Spread—22"   Length—36"

No. 9045—5 Lights

Spread—22"   Length—36"

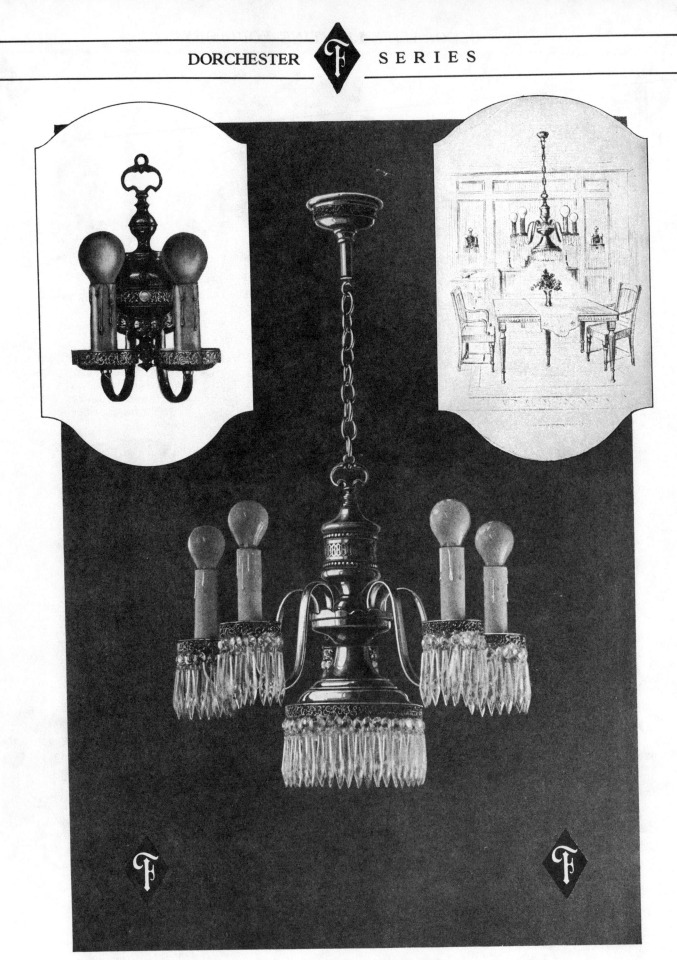

No. 1508—5 x 1 **Lights**

Spread—22"   Length—36"

Dorchester Silver or Antique Brass

FIXTURES
ON THIS PAGE
CAN BE
FURNISHED ONLY
IN ALCAZAR
FINISH

No. 1412—2 Lights

Diameter—8″   Length—36″

6 x 8 Amber Crackled Cylinder

Wired Complete with Glass **$17.00**

Bare with Glass **$16.00**

No. 1415

Spread—10″   Length—36″

4″ x 9″ Acorn

Wired Comp. with Glass **$11.15**

COMPLETE STOCKS—PROMPT SHIPMENTS

MANUFACTURED BY

The Frankelite Co.

MANUFACTURERS OF

Diamond Ⓕ Lighting Equipment

5016 Woodland Avenue   Cleveland, Ohio

Distributed By

ALCAZAR: A finish that is strictly in keeping with the character of this line, giving to every piece, individual charm and beauty. Pleasingly new and different. A beautiful combination of hammered effect copper, richly antiqued with a soft shade of green to which is added a tint of cherry red and russet brown. Harmoniously blending together, producing a richly antiqued, fascinating finish that will retain its beauty indefinitely.

No. 505
Height—70″ overall
Each Complete **$36.50**

Fine Workmanship, Quality Material, Service, Attractive Prices

No. 3090/5
Iron Flush Exit
Depth—3″   Insert Box—6″ x 12″
Face—8″ x 14″ Height of Letters—5″
Black Finish   *Code: Revoke*

No. 3085/5
Iron Surface Exit
Depth—3″   Face—6″ x 12″
Height of Letters—5″ Black Finish
Code: Retreat

No. 3080/5
Iron Triangular Exit
Length—17″   Extension—9″
Panel Size—6″ x 12″
Height of Letters—5″
Black Finish
Code: Respond

No. 3035
Length Overall—40″
Diameter—12″ Depth Lantern—19″
Material: Hammered Copper
Code: Rector

No. 3040
Extension—13½″   Height Overall—21½″
Diameter—12″   Wall Plate—6″ x 14″
Material: Hammered Copper
Code: Refined

No. 3025
Length Overall—18″ Canopy Diam.—4¼″
Lantern Depth—9″ Roof Diam.—5¼″
Open Bottom—Equipped with Cross Bar
Material: Hammered Copper
Code: Record

No. 3050
Extends—5¼″ Length Overall 10½″
Wall Plate—4½″ x 8¾″
Open Bottom—Equipped with Sign
Receptacle Adapter
Material: Hammered Copper

No. 3055
Extends—6″   Length Overall—9½″
Wall Plate—4½″ x 8¾″ Open Bottom
Equipped with Sign Receptacle Adapter
Material: Hammered Copper

No. 3030
Length Overall—17″ Canopy Diam.—5″
Lantern Depth—8″ Roof Diam.—6″
Open Bottom and equipped with Cross Bar
Material: Hammered Copper
Code: Recruit

ALL COPPER NUMBERS
OXIDIZED FINISH WITH
AMBER MOSS GLASS

No. 2202—2 Lights
Spread—10"   Diameter—5"
Code: Jam

No. 2205—5 Lights
Spread—21"   Length—24"
Code: Jar

No. 2201
Extends—5½"   Length—12"
Code: Jade

VICTORIA FINISH: A beautiful combination of hammered effect Brass, richly antiqued with a soft shade of russet brown with just a touch of cherry red and blue. Color scheme similar to LaSalle.

VICTORIA SERIES IS MADE OF SOLID BRASS.

VICTORIA SERIES IS MADE OF SOLID BRASS.

No. 2207—5 Lights
Spread—21"   Length—36"
Code: Jest

No. 2216—3 Lights
Spread—8"   Length—36"
Code: Join

No. 2206—5 Lights
Spread—21"   Length—36"
Code: Java

No. 1561

Extends—6"   Back—11" x 5"

Code: Dial

No. 1562

Spread—9"   Back—11" x 5"

Code: Dictum

No. 1565—5 Lights

Spread—20"   Length—20"

Code: Digest

Fixtures shown on this page can be furnished only in Colonial Silver or Colonial Bronze Finish.

No. 1568—5 Lights

Spread—20"   Length—36

Code: Dim

No. 1566—5 Lights

Spread—20"   Length—36"

Code: Dilate

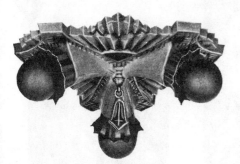

No. 1063—3 Lights
Spread—11½"   Length—4"
Decorated in Ivory and Poly
or Antique Gold
Code: *Anvil*

No. 2631
Extends—3"   Back—8½" x 4½"
Decorated in Ivory and Poly
or Antique Gold
Code: *Cobra*

No. 1062
Spread—12" x 6"   Length—4"
Decorated in Ivory and Poly
or Antique Gold
Code: *Antler*

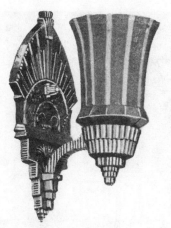

*ATTRACTIVE IN
DESIGN AND FINISH*

No. 2621
Extends—6"   Back—10" x 5½"
Complete with glass shade as shown
Code: *Cream*

No. 2622
Spread—9"   Back—10" x 5½"
Complete with glass shades as shown
Code: *Crew*

*A MODERN CREATION
OF DISTINCTION*

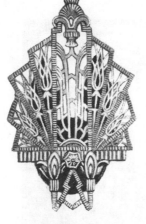

(PATENT APPLIED FOR)

No. 2632—2 Lights
Spread—7"   Length—12"
Code: *Crowd*

DECORATIVE LIGHTING EQUIPMENT

No. 1031

Extends—5¾″   Height—11″

No. 1032

Spread—9″   Diameter—4½″

FINE WORKMANSHIP

We use only the finest wood polished prisms, jewels, and drops.

No. 7060—3 Lights

Spread—16″   Height—16″

Wired Complete with Crystals

Fixtures on this Page Decorated only in English Silver

No. 1060

Beaded Bowl   Diam.—8″

Also **No. 1061**—10″

No. 7057—3 Lights

Spread—16″   Height—36″

Wired Complete with Crystals

No. 7063—1 Light

Spread—10″   Height—10″

Wired Complete with Crystals

No. 7054—7 Lights

Spread—14″   Height—36″

Wired Complete with Crystals

LOUIS XIV

The style of this Period is massive and ornate, but always balanced, and executed with fine dignity, as this royal lamp shows.

ELECTRIC LAMPS

representing all the Periods from the Classic to the Modern are on exhibition at our specially designed show-rooms, Eighth Floor, 11 West 32nd Street, New York. We would call attention to the artistic superiority of these Lamps. Each one possesses decorative individuality that not only gives it value as an æsthetic unit, but that also classes it with some decorative school and so makes it available for use in Period decoration.

ITALIAN RENAISSANCE

Gorgeous and stately, flushed with the colors of grape and sun and sky, this notable work of art stands as a wonderful representation of this Period of the Italian awakening.

THE DUFFNER & KIMBERLY COMPANY
11 WEST 32D STREET NEW YORK

VIKING

The bold and venturesome characteristics of these Sea Rovers are noted in this Electric Lamp, with its shade of rich barbaric colors, adorned with prow-like heads of sea monsters.

ROMAN

The Roman architecture and decoration, with all the imperial desire for imperishable beauty, have all been shown forth in this Electric Lamp.

A

No. 3100—Green.
Code: *Glimpse*

No. 3100—Rose.
Code: *Govern*

No. 3100—Blue.
Code: *Grant*

No. 3105—White Holder with Black, Blue or Green Line Globe.
Code: *Grasp*

No. 3175—Black, Blue or Green Line.
Code: *Port*

No. 3175—Plain White.
Code: *Poor*

No. 3170—W. Conv. Recep. Plain White or Black, Blue or Green Line.
Code: *Plant*

No. 3165—Without Recep. Plain White or Black, Blue or Green Line.
Code: *Pier*

No. 3185—W. Conv. Recep. Plain White or Black, Blue or Green Line. Code: *Pulse*

No. 3180—Without Recep. Plain White or Black, Blue or Green Line. Code: *Pride*

No. 3150—Black Line.
Code: *Panther*

No. 3160—W. Conv. Recep.
Code: *Perish*

No. 3150—Blue Line.
Code: *Parade*

No. 3160—W. Conv. Recep.
Code: *Permit*

No. 3150—Green Line.
Code: *Panel*

No. 3160—W. Conv. Recep.
Code: *Period*

No. 3150—Ivory.
Code: *Peruse*

No. 3160—W. Conv. Recep.
Code: *Patrol*

No. 3150—Green.
Code: *Persist*

No. 3160—W. Conv. Recep.
Code: *Pardon*

No. 3150—Orchid.
Code: *Pervade*

No. 3160—W. Conv. Recep.
Code: *Pause*

No. 3150—Code: *Pedal*

No. 3160—W. Conv. Recep.
Code: *Pad*

Lightolier

569 BROADWAY   NEW YORK CITY

C

6930

HANDEL Lamps are ever in accord with the best in interior decoration. To the richness of Oriental rugs, the dignity of fine old paintings and the gleam of deep mahogany, they bring harmony of color and grace of line not found in ordinary lamps. All glass shades are hand-painted, all metal parts superbly finished. Handel Lamp No. 6930 goes admirably on the library table. No. 6896 is a unique davenport lamp. See Handel Lamps in the shops you visit, or write to us for the name of the nearest dealer. The Handel name is on every lamp.

THE HANDEL COMPANY, Meriden, Conn.

HANDEL Lamps

6896

*Table Lamp
No. 6884*

*Adjustable
Floor Lamp
No. 6893*

Like the handiwork of an "Old Master"

IN all its glowing loveliness, a Handel Lamp makes an irresistible appeal. From base to shade, it is a perfect harmony of line and color—and of material that will forever hold its beauty. The colors are permanent and fadeless and after years of use will be bright as when new. All metal parts are specially treated, resulting in a finish and decoration as permanent as the metal itself.

**THE HANDEL COMPANY
Meriden, Conn.**

HANDEL *Lamps*

Floor lamps; lamps for the boudoir and alongside the easy chair; lamps for the library and for every corner that needs a spot of bright color—all these are made by the Handel master artists and sold by best dealers everywhere. Look for them in the window. The Handel name is on every genuine Handel Lamp.

E

F

No. 7088

HANDEL LAMPS are noteworthy examples of master craftsmanship. Each is designed and made for permanence, to give life-long service as well as to enhance the attractiveness of the room it adorns. The lamp pictured above illustrates the unusual in decorative treatment, the rare artistry of line and fine balance between shade and standard that make Handel Lamps so distinctive and so desirable.

All Handel shades are painted by hand with colors that are fadeless. All standards and metal parts specially treated to afford a decorative finish as enduring as the metal itself.

Floor lamps, electroliers, pendants, wall sconces, boudoir lamps and torcheres—a few of which are shown—provide a Handel Lamp for every purpose, for every room. They, or other styles and designs, may be purchased at stores of the better sort. Ask to see them. The name Handel is on every lamp.

HANDEL Lamps

H

ELECTROLIERS

✞ No. 5025   Height 27 Inches
Turinese or Flemish Bronze
Each ....

✞ No. 5020   Height 23 Inches
Turinese or Flemish Bronze
Each ....

✞ No. 5030   Height 26 Inches
Turinese or Flemish Bronze
Each ....

✞ The Above Electroliers are Wired Complete, and Above Prices Include Electric Lamps, 6 Feet Cord and Attachment Plug Complete.

ELECTROLIERS

All Lamps Fitted with Wire Cord and Sockets for Bulbs

No. 97643   No. 97644   No. 97645   No. 97638

Per doz.

97643   Cast metal base, antique finished, one burner with patent socket, cathedral glass shade set in metal frame to match base, beaded fringe, height 23 inches, size of shade 11½ inches ....

97644   Heavy metal base, hammered steel finished, one burner, patent socket, cathedral glass set in hammered antique steel, size of shade 8½ inches, height 19½ inches...

97645   Solid metal base, old brass finish, one burner with patent socket, cathedral glass set in old brass frames, size 12 inches with beaded fringe, height 21 inches ....

97638   Solid metal base, brushed brass finished, 2 burners, patent sockets, shaded cathedral glass set in solid metal frame, diameter of shade 13 inches, height 18 inches.....

No. 97646   No. 97639   No. 97647   No. 97634

Per doz.

97646   Brushed brass base, one burner with patent socket, two contrasting colored glass, set in brushed brass frame to match base, size 12½ inches, height 20 inches ....

97639   Solid brass base, old brass finish, one burner with patent socket, adjustable shade, cathedral glass, assorted colors, specially adapted for library table, length of shade 9 inches, height 12½ inches ....

97647   Brushed brass base, one burner with patent socket, cathedral glass, set in brushed brass shade, size 12 inches, height 22 inches ....

97634   Onyx and solid metal gold plated base, one burner with patent socket, removable glass shade with conventional designs in assorted subdued colors, diameter of shade 12 inches, height of lamp 24 inches ....

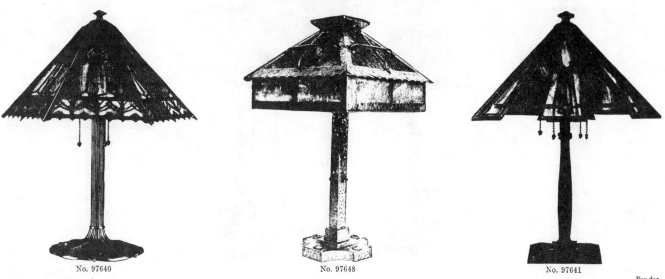

No. 97640   No. 97648   No. 97641

Per doz.

97640 Solid brass base, diameter 9 inches, finished in antique brass, 3 burners with patent sockets, extra fine shaded cathedral glass, set in solid metal, handsome designed metal over glass, diameter of shade 18 inches, height 21½ inches ....

97648 Solid metal base, old hammered steel finished, 2 burners with patent sockets, antique shade to match, set with cathedral glass, size 13 inches, height 24 inches....

97641 Extra heavy base, antique finished, 4 burners with patent socket, assorted colored handsomely shaded cathedral glass, with conventional design metal overlay, size 16 inches, height 22½ inches ....

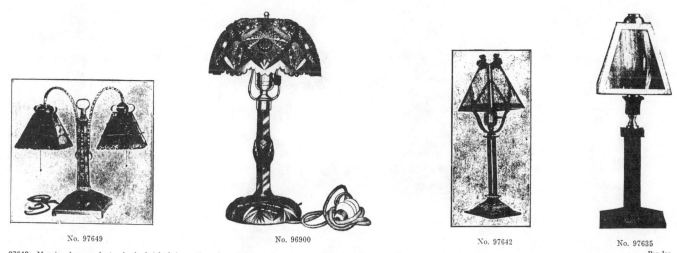

No. 97649   No. 96900   No. 97642   No. 97635

Per doz.

97649 Massive base and standard, finished in antique hammered steel, 2 shades, with shaded cathedral glass set in metal to match, size 7 inches, 2 burners with patent sockets, height 19½ inches, width 19 inches ....

97642 Metal base antique steel finish, 1 burner with patent socket, cathedral glass set in metal frame to match base, size 5 inches, height 17½ inches ....

97635 Brushed brass finished base, 1 burner, shaded cathedral glass set in metal frame to match, size 5½ inches, height 14½ inches ....

96900 Fine cut glass standard and shade, diameter of shade 12 inches, fitted attachments ....

Per doz.

97637 Metal base, old brass finish, handsomely embossed, frosted glass globe, 1 burner with patent socket, height 9¾ inches ....

97636 Metal, old brass finish, 1 burner with patent socket, frosted glass bulb, length 8 inches, height 8¾ inches ....

No. 97637   No. 97636

ELECTROLIERS

No. 80464   No. 80463   No. 80465   No. 80466

No.
80464   Brass, antique finish, glass shade with floral design in natural colors on light background, one burner, fitted with pull chain, 6 feet silk cord and electric plug, height 16 inches

80463   Metal base and standard, dark verde finish, shade is combination of green art glass and metal in dark verde, green fringe, chain pull and 6 feet of silk cord with socket...

No.
80465   Brass, Roman gold finish, handsome glass shade ornamented with beautiful floral design in natural colors, green background, 2 burners, fitted with pull chain, 6 feet silk cord and electric plug, height 16 inches...

80466   Brass base and beautiful openwork brass shade over glass in beautiful colors, 2 burners, fitted with pull chain, 6 feet silk cord and electric plug, height 16 inches....

No. 80468   No. 80462   No. 80460   No. 70528

No.
80468   Old brass standard engraved, beautiful shade, delicate green background, handsome floral design in most delicate tints, one light fitted with electric plug and 8 feet best silk cord, height 21 inches, diameter of shade 12 inches...

80461   Full French bronze, "The Reaper" full height 24 inches, height of figure exclusive of base 14½ inches, 2 burners, fitted with 6 feet of silk cord and electric plug......

80462   Full French bronze, "The Sower" full height 25 inches, height of figure exclusive of base 15½ inches, 2 burners, fitted with six feet silk cord and electric plugs.......

No.
80460   Brass, brass shade, Roman gold finish, fancy glass ornaments on side of shade, large glass ornament on top, fancy bead fringe, fitted with chain pull, electric socket for lamp, wired, 6 feet of silk cord and electric plug, height 18 inches.....................

70528   Romanesque and Colonial brass, 8-inch Murano shade with ivory top, green and opal band, and ruby border, chain pull, electric socket and one 10 c. p. round bulb lamp, wired, six feet silk cord and electric plug.................................

No.
70530   Colonial brass, 12-inch art glass shade, green and opal glass, with 5-inch green bead fringe, chain pull, electric sockets for two 16 c. p. lamps, wired, 6 feet silk cord and electric plug, height to shade 13 inches, extreme height 19 inches...........

80467   Old brass standard handsomely engraved, opalescent shade, most delicate green background, hand-painted design of roses and lace, also delicate lace decoration, 2 lights, height 22½ inches, diameter of shade at top 9½ inches, at bottom 15½ inches, fitted with electric plug and 8 feet best silk cord.................................

No. 70530   No. 80467

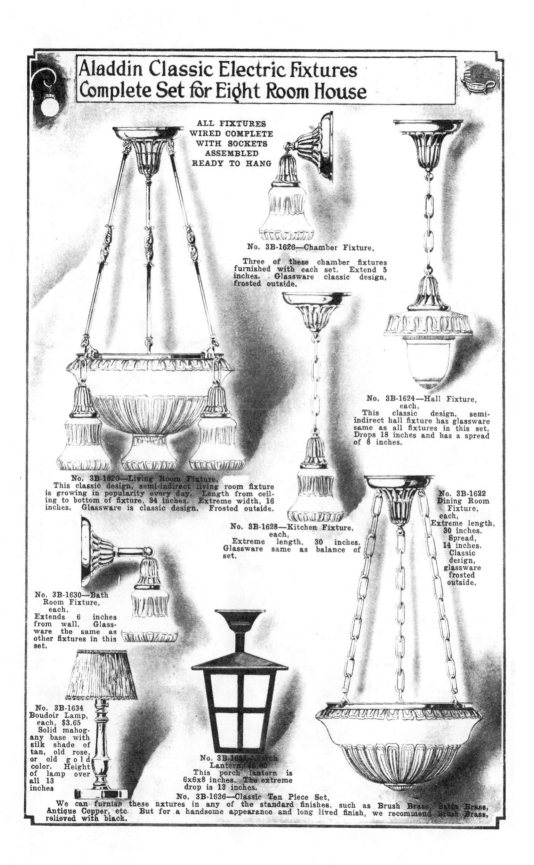

Aladdin Classic Electric Fixtures
Complete Set for Eight Room House

**ALL FIXTURES
WIRED COMPLETE
WITH SOCKETS
ASSEMBLED
READY TO HANG**

No. 3B-1626—Chamber Fixture,

Three of these chamber fixtures furnished with each set. Extend 5 inches. Glassware classic design, frosted outside.

No. 3B-1624—Hall Fixture, each,
This classic design, semi-indirect hall fixture has glassware same as all fixtures in this set. Drops 18 inches and has a spread of 8 inches.

No. 3B-1620—Living Room Fixture,
This classic design, semi-indirect living room fixture is growing in popularity every day. Length from ceiling to bottom of fixture, 34 inches. Extreme width, 16 inches. Glassware is classic design. Frosted outside.

No. 3B-1622 Dining Room Fixture, each,
Extreme length, 30 inches. Spread, 14 inches. Classic design, glassware frosted outside.

No. 3B-1628—Kitchen Fixture, each,
Extreme length, 30 inches. Glassware same as balance of set.

No. 3B-1630—Bath Room Fixture, each,
Extends 6 inches from wall. Glassware the same as other fixtures in this set.

No. 3B-1634 Boudoir Lamp, each, $3.65
Solid mahogany base with silk shade of tan, old rose, or old gold color. Height of lamp over all 13 inches

No. 3B-1632—Porch Lantern, $5.00
This porch lantern is 6x6x8 inches. The extreme drop is 13 inches.

No. 3B-1636—Classic Ten Piece Set,

We can furnish these fixtures in any of the standard finishes, such as Brush Brass, Satin Brass, Antique Copper, etc. But for a handsome appearance and long lived finish, we recommend Brush Brass, relieved with black.

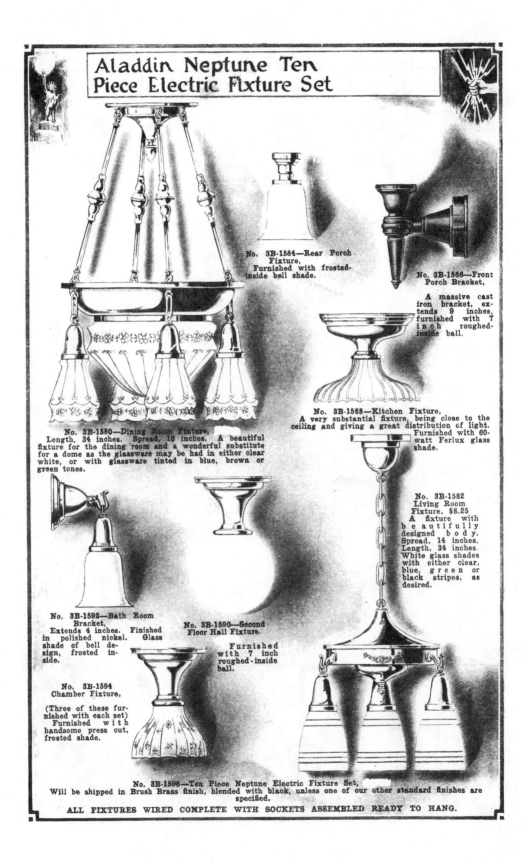

Aladdin Neptune Ten Piece Electric Fixture Set

No. 3B-1584—Rear Porch Fixture, Furnished with frosted-inside bell shade.

No. 3B-1586—Front Porch Bracket,

A massive cast iron bracket, extends 9 inches, furnished with 7 inch roughed-inside ball.

No. 3B-1580—Dining Room Fixture, Length, 34 inches. Spread, 16 inches. A beautiful fixture for the dining room and a wonderful substitute for a dome as the glassware may be had in either clear white, or with glassware tinted in blue, brown or green tones.

No. 3B-1588—Kitchen Fixture, A very substantial fixture, being close to the ceiling and giving a great distribution of light. Furnished with 60-watt Ferlux glass shade.

No. 3B-1582 Living Room Fixture, $8.25 A fixture with beautifully designed body. Spread, 14 inches. Length, 34 inches. White glass shades with either clear, blue, green or black stripes, as desired.

No. 3B-1592—Bath Room Bracket, Extends 4 inches. Finished in polished nickel. Finished Glass shade of bell design, frosted inside.

No. 3B-1590—Second Floor Hall Fixture.

Furnished with 7 inch roughed-inside ball.

No. 3B-1594 Chamber Fixture,

(Three of these furnished with each set) Furnished with handsome press cut, frosted shade.

No. 3B-1596—Ten Piece Neptune Electric Fixture Set, Will be shipped in Brush Brass finish, blended with black, unless one of our other standard finishes are specified.

ALL FIXTURES WIRED COMPLETE WITH SOCKETS ASSEMBLED READY TO HANG.

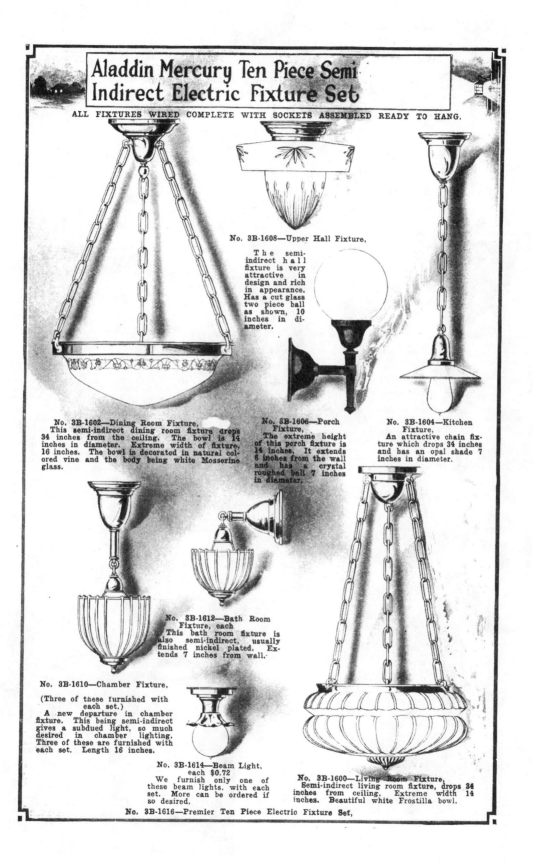

Aladdin Mercury Ten Piece Semi Indirect Electric Fixture Set

ALL FIXTURES WIRED COMPLETE WITH SOCKETS ASSEMBLED READY TO HANG.

No. 3B-1608—Upper Hall Fixture.

The semi-indirect hall fixture is very attractive in design and rich in appearance. Has a cut glass two piece ball as shown, 10 inches in diameter.

No. 3B-1602—Dining Room Fixture.
This semi-indirect dining room fixture drops 34 inches from the ceiling. The bowl is 14 inches in diameter. Extreme width of fixture, 16 inches. The bowl is decorated in natural colored vine and the body being white Mosserine glass.

No. 3B-1606—Porch Fixture.
The extreme height of this porch fixture is 14 inches. It extends 8 inches from the wall and has a crystal roughed ball 7 inches in diameter.

No. 3B-1604—Kitchen Fixture.
An attractive chain fixture which drops 34 inches and has an opal shade 7 inches in diameter.

No. 3B-1612—Bath Room Fixture, each
This bath room fixture is also semi-indirect, usually finished nickel plated. Extends 7 inches from wall.

No. 3B-1610—Chamber Fixture.
(Three of these furnished with each set.)
A new departure in chamber fixture. This being semi-indirect gives a subdued light, so much desired in chamber lighting. Three of these are furnished with each set. Length 16 inches.

No. 3B-1614—Beam Light, each $0.72
We furnish only one of these beam lights, with each set. More can be ordered if so desired.

No. 3B-1600—Living Room Fixture.
Semi-indirect living room fixture, drops 34 inches from ceiling. Extreme width 14 inches. Beautiful white Frostilla bowl.

No. 3B-1616—Premier Ten Piece Electric Fixture Set.

Ivanhoe Fixtures for Commercial Lighting

The
Louvre

The
Beehive

The
Dualite

The Louvre—A modern indirect lighting unit, distinctive and graceful in appearance. Combining utility with beauty, it provides a wide distribution of reflected light through the use of Ivanite finished aluminum inside reflecting surface on the top louvre. Additional light filtering through the lower louvres softly illuminates the entire fixture. By inserting a color screen in the bottom cap, the color of the fixture may be changed as desired to blend with room decorations. When color screen is used the efficiency of the unit is not decreased. The bottom knob is softly luminous.

Color screens are included with the Louvre in red, green and amber colors. Color screens will be packed with fixture.

The Beehive—An indirect ball type unit—an engineered lighting unit—not a common ball globe. Modern treatment and efficiency demands the new and simple construction to provide ease of installation and maintenance and consideration of brightness to eliminate eye strain so often forgotten in the design of modern lighting equipment.

The glassware is satin finish inside. It is securely and rigidly locked to the stem by a saddle strap, fitter cover and lock nut. The aluminum bottom is Ivanite finished inside and is held in place by three spring clips, easily removed for relamping, and is of sufficient height to shield the lamp filament from the eye.

The Dualite—A practical lighting unit designed on semi-indirect principles to meet a wide range of lighting requirements. The light is entirely diffused and is of a soft, uniform quality. The greater percentage of light is reflected from the ceiling. The direct rays filtering through the glass bottom bowl are free from all glare. Small aluminum deflector supplied with each unit serves a double purpose—increases efficiency—reduces downward component of light, making the glass bowl luminous. This deflector may or may not be used depending upon service desired.

The deflector has an inner reflecting surface of white porcelain enamel. The outside of deflector is of Ivre Porcelain Enamel. Easy to clean—durable.

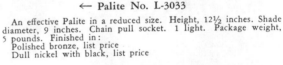

← Palite No. L-3033

An effective Palite in a reduced size. Height, 12½ inches. Shade diameter, 9 inches. Chain pull socket. 1 light. Package weight, 5 pounds. Finished in:
Polished bronze, list price
Dull nickel with black, list price

Palite No. L-3023 →

Effective as a desk or occasional lamp. Combines decorative value with utility. Height, 15 inches. Shade diameter, 11 inches. Chain pull socket. 1 light. Package weight, 6 pounds. Finished in:
Polished bronze, list price
Dull Nickel with Black, list price

Ivanhoe Duo-Purpose Type S-1 Ultra-Violet Lighting Fixtures

Indirect-Type—Catalog No. 2113
Satin Aluminum Finish........List Price
   Diam. Reflector, 20 in.; Diam. Ceiling
Reflector, 36 in.; Length, 16 in.

Catalog No. 2112
   As above, without Aluminum Ceiling Reflector.
Satin Aluminum Finish........List Price
   Diam. Canopy, 13½ in.; Length, 16 in.

These fixtures were designed for use in private offices, directors' rooms, libraries, auditoriums, reception rooms, solariums, beauty parlors, barber shops, restaurants, banks, exclusive shops, studios, homes, and similar places.

Direct-Semi-Indirect Type
With Silk Shade, Catalog No. 2100
(As illustrated)

Catalog No.	Bronze Finish	Shaded Nickel Finish
2100 List Price..................		
With Parchment Shade		
2101 List Price..................		
With Mica Shade		
2102 List Price..................		
With Parchment Shade with Metal Overlay		
2103 List Price		
With Mica Shade with Metal Overlay		
2104 List Price		

Diameter, 23 inches; Length, 36 inches.

Direct-Semi-Indirect Type
Catalog No. 2114
Satin Aluminum Finish........List Price
   Diam., 19 in.; Diam. Canopy, 9 in.;
Length, 42 in.

Direct-Semi-Indirect Type
With Parchment Shade, Catalog No. 2109
(As illustrated)

Catalog No.	Bronze Finish	Shaded Nickel Finish
2109 List Price..................		
With Mica Shade		
2110 List Price		
With Parchment Shade without Metal Overlay		
2107 List Price		
With Silk Shade without Metal Overlay		
2106 List Price		
With Mica Shade without Metal Overlay		
2108 List Price		

Diameter, 23 inches; Length, 16 inches.

Type S-1 Duo-Purpose lighting fixtures have two circuits. One circuit is for the direct lighting ultra-violet radiation and employs one Type S-1 Mazda Sunlight Lamp. In the semi-indirect type fixtures it is only necessary to use this circuit for relatively short periods of time to obtain the desirable health benefit of ultra-violet light. With the indirect type this circuit may be used, in most cases, all day without the likelihood of sunburn in that the ultra-violet illumination is in reflected rather than direct form. The other circuit employs ordinary Mazda lamps and is designed to be used for long periods of continued visual illumination, in the same manner that any other direct or semi-indirect lighting fixture is used.

Outstanding Features of Ivanhoe Duo-Purpose Fixtures—Efficient Ultra-Violet Utilization. Scientific Visible Radiation Design. Freedom from Glare. Flexibility in Indirect Component. Pleasing Color Quality of Light with Both Circuits. Entire Fixture Softly Luminous with Either Circuit. Variety of Decorations and Finishes. Ease of Installation and Relamping.

It is not necessary to obtain sunburn to receive ultra-violet benefits. It is believed that several hours' daily usage of this Indirect Fixture, and perhaps even shorter periods, will be distinctly beneficial to the average user from the health-maintenance standpoint.

Circuits: Fixtures are wired for two circuits; one Type S-1 Mazda Sunlight Lamp is used on one circuit, and in the Direct-Semi-Indirect type six 40, 60, 75 or 100-watt Mazda lamps on the other circuit. In the Indirect type five 75 or 100-watt Mazda lamps are used in the other circuit. Fixtures should be installed on a separate service circuit. Each circuit should have individual switch control. Switch control may be either by a twin wall type switch of 10 ampere capacity, or by a 10 ampere two-way pull switch installed on ceiling. Installation of these fixtures usually requires additional wiring.

Service: Duo-Purpose fixtures operate only on 110-120 volt, 60 cycle, A.C. Service. The transformer is embodied in the fixture.

Note: List prices shown are for fixtures wired complete. Lamp bulbs are not included. Silk, parchment and mica shades are packed separately.

Faries Adjustable Portables

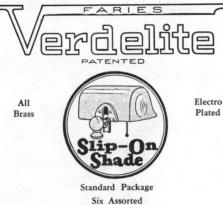

FARIES

Verdelite

PATENTED

All
Brass

Electro
Plated

Slip-On
Shade

Standard Package
Six Assorted

THE ORIGINAL GREEN SHADE LAMP

The Verdelite shade is made of two compositions of glass fused into one solid piece. Outside is cool, restful green — inside is soft, white opal. By using a Mazda inside frosted blue daylight bulb, this shade gives the true daylight effect.

NO COMPLICATED PARTS TO ADJUST

Verdelite lamps offer many exclusive patented features. Shades are instantly slipped on or off special holder by hand, and attached securely without springs, hinges, or set screws. Easy to clean and adjust — less likely to break than other makes.

No. 3241

Fancy solid brass swing portable with vertical adjustment. Admirably adapted for use on pianos or roll top desks. Extends 9½ in. For 25 to 60 watt lamps. Wired with 9 ft. silk cord, pull chain socket and plug.

Brushed Brass; Black Relief......
Statuary Bronze; Green Relief

Glass
Shade
No. 3600M

No. 3211-M

Deluxe model solid brass portable — graceful, strong, and unusually attractive. For executives' desks, writing desks, and in the home. Height 13½ in. to center of shade. Equipped with beautifully etched green glass shade. For 25 to 60 watt lamp.

Wired with 9 ft. silk cord, pull chain socket and plug.

Statuary Bronze ....

No. 3260

Fancy solid brass swing portable with vertical adjustment. Admirably adapted for use on pianos or roll top desks. Extends 10 in. For 25 to 60 watt lamps. Wired with 9 ft. silk cord, pull chain socket and plug.

Brushed Brass; Black Relief....
Statuary Bronze; Green Relief

No. 3230

The standard design in a solid brass portable. Adaptable for flat top desks and tables. Height 15 in. to center of shade. For 25 to 60 watt lamps. Wired with 9 ft. silk cord, pull chain socket and plug.

Brushed Brass; Black Relief.....
Statuary Bronze; Green Relief

VERDELITE GLASS SHADES

Regular Verdelite glass shade shown above is No. 3134 described on page 133.

Shade No. 3600M shown on No. 3211M ....

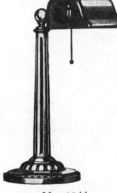

No. 3544

Fancy solid brass portable with round embossed base and fluted column. Height 15 in. to center of shade. For 25 to 60 watt lamps. Wired with 9 ft. silk cord, pull chain socket and plug.

Brushed Brass; Black Relief....
Statuary Bronze ....

Faries Adjustable Portables

FARIES
Verdelite
PATENTED

All Brass   Electro Plated

Slip-On Shade

Standard Package
Six Assorted

Glass Shade No. 3600A

No. 3230A

Fancy solid brass portable with adjustable and detachable green crackled glass shade having etched scroll on opal border. Height 15 inches to center of shade. For 25 to 60 watt lamps. Wired with 9 ft. silk cord, pull socket and plug.

Statuary Bronze; Green Relief..

Glass Shade No. 3600B

No. 3544B

Ornamental seamless brass portable with adjustable and detachable brown crackled glass shade having etched scroll on opal border. Height 15 inches to center of shade. For 25 to 60 watt lamps. Base 7½ inches. Wired with 9 ft. silk cord, pull socket and plug.

Brushed Brass; Black Relief....

Glass Shade No. 3600M

No. 920M

New low type modernistic brass portable with adjustable and detachable beautifully etched green glass shade. Base 7 x 5¼ inches, channeled across front for pen or pencil. Height only 11½ inches overall. For 25 to 60 watt lamps. Wired with 9 ft. cord, pull socket and plug.

Polished Chrome and Black......

No. C920

Exactly same as No. 920M except equipped with adjustable and detachable Chrome plated metal shade as shown on portable No. C921 at right.

Polished Chrome and Black......

No. 3297

A graceful all brass portable of modern design with adjustable and detachable plain green glass shade. Base 6 inch diameter. Height to center of shade 12 inches. For 25 to 60 watt lamps. Wired with 9 ft. cord, pull socket and plug.

Polished Chrome ....

No. 1563

Exactly same as No. 3297 except equipped with adjustable and detachable Chrome plated metal shade as shown on portable No. C921 at right.

Polished Chrome ....

SHADES ONLY LESS HOLDER

No. 3600A	....
No. 3600B	....
No. 3600M	....
No. 3201	....

Interchangeable on any Verdelite portable at difference in prices shown above.

Metal Shade No. 3201

No. C921

The new, pyramid type of modernistic portable with adjustable and detachable metal shade. Low type, only 11½ inches high. Base 7 x 4½ inches, channeled across front for pen or pencil. For 25 to 60 watt lamps. Wired with 9 ft. cord, pull socket and plug.

Polished Chrome and Black......

No. 921M

Exactly same as No. C921 except equipped with adjustable and detachable beautifully etched green glass shade as shown on portable No. 920M at left.

Polished Chrome and Black....

Faries Adjustable Portables

Standard Package
Six Assorted

No. 3243

Plain solid brass swing portable, vertical adjustment with detachable green glass shade. Admirably adapted for roll top desk or piano. Extends 10 inches. For 25 to 60 watt lamp. Wired with 9 feet silk cord, pull socket and plug.

Brushed Brass ....

Statuary Bronze ....

No. 3258

A practical and economical wall bracket with adjustable and detachable green glass shade. Extends 9 inches to center of shade. For 25 to 60 watt lamp. Wired complete with fixture wire and pull chain socket ready to attach.

Brushed Brass ....

Statuary Bronze ....

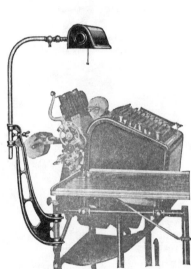

No. 3248

A solid brass plain portable equipped with adjustable and detachable green glass shade. For flat top desks. Heavy brass base. Height 15 inches to center of shade. For 25 to 60 watt lamps. Wired with 9 feet silk cord, pull socket and plug.

Brushed Brass ....

Statuary Bronze ....

No. 3238

Adding or bookkeeping machine bracket. Cast arm clamps to post. Heavy brass pipe stem extends 13 inches from back to center of shade and from 20 to 35 inches in height. Double adjustable joint behind shade. Equipped with adjustable and detachable green glass shade. Wired with 9 feet cord, pull socket and plug.

Statuary Bronze Stem, Bracket
Black Enamel ....
Specify whether for round or square post.

No. 3218

For stenographic desks. Base is felted and attaches to desk with clamp. Arm swings horizontally and extends 18 to 24 inches. Base 4½ inches square. Equipped with adjustable and detachable green glass shade. Wired with 9 feet silk cord, pull socket and plug.

Brushed Brass ....

Statuary Bronze ....

No. 3134 Shade Only

Verdelite plain green glass shade has white opal reflecting surface inside. Slotted in back for special holder. Fits all Verdelite portables.
S.P. 12. Price each, less holder ....

No. 3135 Shade Holder Only

Two piece cast clamp holder, fits all Verdelite glass shades. Complete with screws. Specify finish desired. S.P. 12, Price each ....

Faries Adjustable Portables

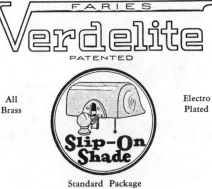

FARIES
Verdelite
PATENTED

All Brass

Electro Plated

Slip-On Shade

Standard Package
Six Assorted

THE ORIGINAL GREEN SHADE LAMP

The Verdelite shade is made of two compositions of glass fused into one solid piece. Outside is cool, restful green — inside is soft, white opal. By using a Mazda inside frosted blue daylight bulb, this shade gives the true daylight effect.

NO COMPLICATED PARTS TO ADJUST

Verdelite lamps offer many exclusive patented features. Shades are instantly slipped on or off special holder by hand, and attached securely without springs, hinges, or set screws. Easy to clean and adjust — less likely to break than other makes.

No. 3241

Fancy solid brass swing portable with vertical adjustment. Admirably adapted for use on pianos or roll top desks. Extends 9½ in. For 25 to 60 watt lamps. Wired with 9 ft. silk cord, pull chain socket and plug.

Brushed Brass; Black Relief......
Statuary Bronze; Green Relief

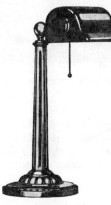

Glass Shade No. 3600M

No. 3211-M

Deluxe model solid brass portable — graceful, strong, and unusually attractive. For executives' desks, writing desks, and in the home. Height 13½ in. to center of shade. Equipped with beautifully etched green glass shade. For 25 to 60 watt lamp.

Wired with 9 ft. silk cord, pull chain socket and plug.

Statuary Bronze ....

No. 3260

Fancy solid brass swing portable with vertical adjustment. Admirably adapted for use on pianos or roll top desks. Extends 10 in. For 25 to 60 watt lamps. Wired with 9 ft. silk cord, pull chain socket and plug.

Brushed Brass; Black Relief....
Statuary Bronze; Green Relief

No. 3230

The standard design in a solid brass portable. Adaptable for flat top desks and tables. Height 15 in. to center of shade. For 25 to 60 watt lamps. Wired with 9 ft. silk cord, pull chain socket and plug.

Brushed Brass; Black Relief....
Statuary Bronze; Green Relief

VERDELITE GLASS SHADES

Regular Verdelite glass shade shown above is No. 3134 described on page 133.

Shade No. 3600M shown on No. 3211M ....

No. 3544

Fancy solid brass portable with round embossed base and fluted column. Height 15 in. to center of shade. For 25 to 60 watt lamps. Wired with 9 ft. silk cord, pull chain socket and plug.

Brushed Brass; Black Relief....
Statuary Bronze ....

No. 3-674
Salmon and White

No. 3-672
Gold and Black

No. 3-673
Blue Bronze

No. 3-642
Russet and Green

No. 3-671
Ivory and Brown

No. 3-623
Gold Brown

Jeannette Lamps — Hand Decorated and Colors Fired In

GROUP 30—BOUDOIR LAMPS

Diameter of shades 8 inches. Stands are equipped with push through sockets, five feet of silk cord and two-piece plugs. Height 13¾ inches. Packed two dozen shades in a shipping carton (four of each lamp in the group). Stands packed in a separate carton.

28H203. Poly-chrome Parrot Lamp in the natural coloring with dainty two-tone braid-trimmed Silk Shade with 5 feet of cord and plug. About 17 inches high...... **$2.98**

28 H 204. Colonial Lady design Boudoir Lamp in old Ivory finish. with natural tinted rosebud trimming. Complete with handsome matching shade. Price **$2.98**

28H211. Buffet or desk Torchiere in rich old gold colorings with **ivory** color fancy candle effect designed to hold electric bulb. Bulb is not included. Complete, with 5 feet of cord and plug. Price. **$1.98** Postage 12¢ extra.

Boudoir Lamp

One of the daintiest lamps that ever shed a soft, pleasing glow over any boudoir setting. Thirteen and one-half in. high; shade 7½ in. diameter. In Patina Brass, Old Ivory or Florentine Relief, each with a shade that harmonizes perfectly. A dream of a present to give *her* for Christmas.

L-2653

70

No. 4-678
Black and Gold

No. 4-676
Gold and Brown

No. 4-677
Ivory and Brown

No. 4-680
Blue Bronze

No. 4-675
Salmon and Pink

No. 4-679
Russet Green

Jeannette Lamps — Hand Decorated and Colors Fired In

GROUP 40—BOUDOIR LAMPS

Diameter of shades 7 inches. Stands are equipped with push through sockets, five feet of silk cord and two-piece plugs. Height 13¾ inches. Packed two dozen shades in a shipping carton (four of each lamp in the group). Stands are packed in a separate carton.

DENZAR—The Unit of Day Brightness

Pendant Types

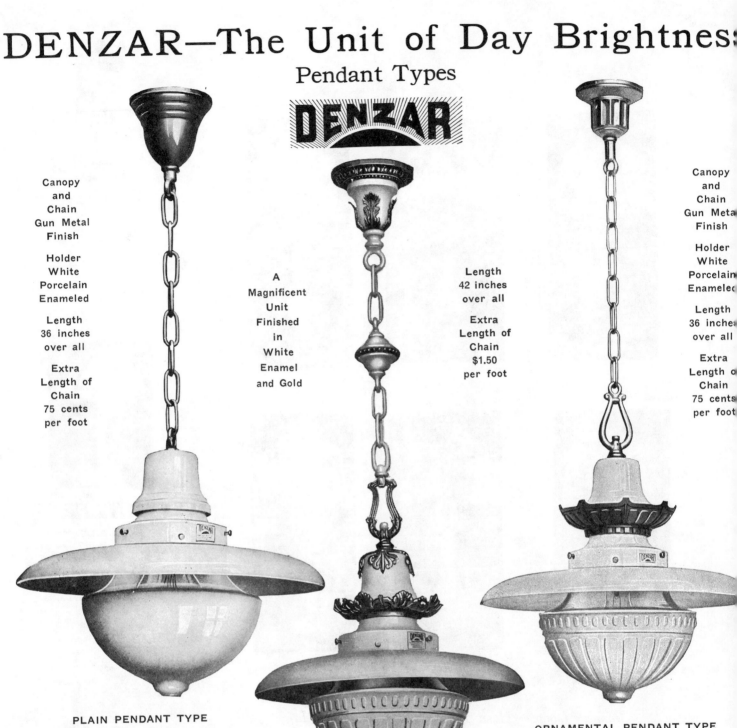

Canopy
and
Chain
Gun Metal
Finish

Holder
White
Porcelain
Enameled

Length
36 inches
over all

Extra
Length of
Chain
75 cents
per foot

A
Magnificent
Unit
Finished
in
White
Enamel
and Gold

Length
42 inches
over all

Extra
Length of
Chain
$1.50
per foot

Canopy
and
Chain
Gun Metal
Finish

Holder
White
Porcelain
Enameled

Length
36 inches
over all

Extra
Length of
Chain
75 cents
per foot

PLAIN PENDANT TYPE

No. Junior—100-150 watt size. 14-inch reflecting dome, 8-inch Denzar bowl, medium screw base socket.   List price,

No. 1—200 watt size. 17-inch reflecting dome, 10-inch Denzar bowl and medium screw base socket.   List price,

No. 1½—300 watt size. 17-inch reflecting dome, 10-inch Denzar bowl and mogul socket.   List price,

No. 2—500 watt size. 20-inch reflecting dome, 12-inch Denzar bowl, mogul socket.   List price,

With Ornamental Bowl

No. 600—100-150 watt size. 14-inch reflecting dome, 8-inch ornamental Denzar bowl, medium screw base socket.   List price,

DENZAR DE LUXE

No. 1000—100-150 watt size. 14-inch reflecting dome, 8-inch ornamental Denzar bowl, medium screw base socket.   List price,

ORNAMENTAL PENDANT TYPE

No. 800—100-150 watt size. 14-inch reflecting dome, 8-inch ornamental Denzar bowl, medium screw base socket.   List price,

No. 801—200 watt size. 17-inch reflecting dome, 10-inch ornamental Denzar bowl, medium screw base socket.   List price,

No. 801½—300 watt size. 17-inch reflecting dome, 10-inch ornamental Denzar bowl, mogul socket.   List price,

No. 802—500 watt size. 20-inch reflecting dome, 12-inch ornamental Denzar bowl, mogul socket.   List price,

With Plain Bowl

No. 300—100-150 watt size. 14-inch reflecting dome, 8-inch Denzar bowl, medium screw base socket.   List price,

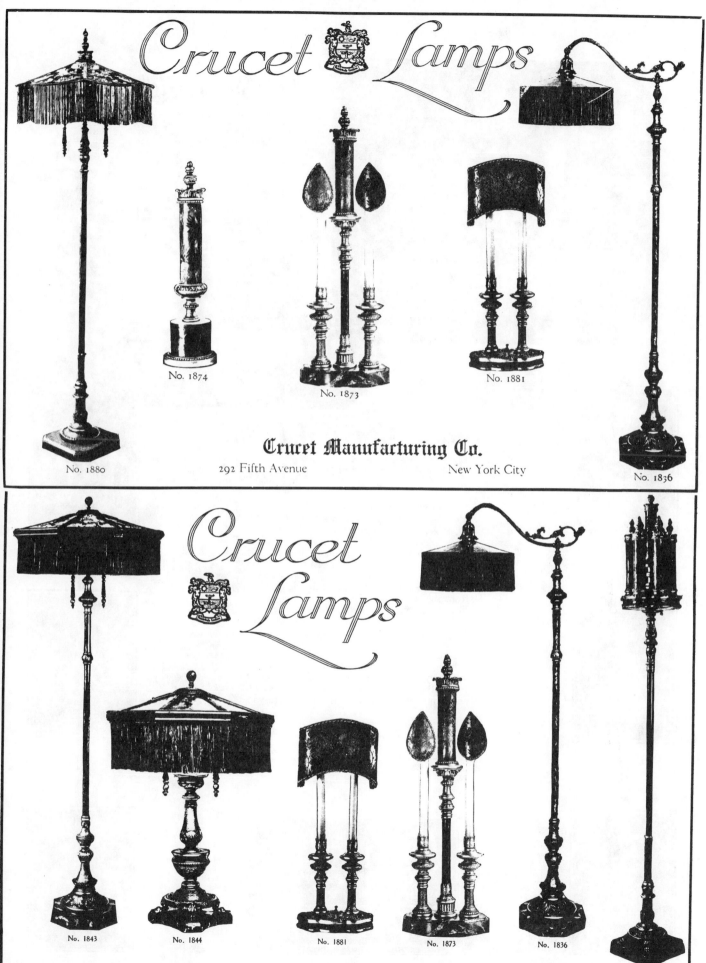

Crucet Lamps

No. 1874

No. 1873

No. 1881

Crucet Manufacturing Co.

292 Fifth Avenue   New York City

No. 1880

No. 1836

Crucet Lamps

No. 1843   No. 1844   No. 1881   No. 1873   No. 1836

No. 1888

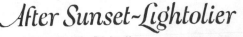

After Sunset—Lightolier

Adam Candle Lightolier

Chippendale Corona Lightolier

Here Are
Two Ideal Ways of Lighting
Your Dining Room

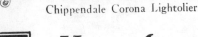

This Louis XVI Corona Lightolier, 72 inches long, with 22-inch shade, is a very beautiful dining room fixture—finished in gold or silver plate.

This exquisite Chippendale Candle Lightolier, 36 inches long and 18 inches spread, is the very acme of lighting fixture grace —finished in gold or silver plate, or Roman gold.

The Corona Lightolier, without glare, throws a flood of brilliant light downward onto the dining table, enhancing the beauty of silver and linen. At the same time, through its patented diffusing bowl, under the shade, it spreads a diffused light throughout the rest of the room.

The Candle Lightolier, when at the right height from the table and equipped with the proper silk shades, brightly lights every part of the dining room, while reflecting down upon the table a pleasant and sufficient illumination.

Just as Lightolier Company has solved the lighting problems of the dining room, so has it equally well solved the lighting problems of every other room in the home—both in fixtures and lamps—and at a minimum of cost to you.

This Adam Corona Lightolier, 72 inches long, with 22-inch shade, is a very unusually graceful design—finished in gold or silver plate, or Roman gold.

This is a Candle Lightolier that is so different that Lightolier is very proud of it. Length 36 inches, spread 16 inches, with five lights—finished in a new exclusively Lightolier finish.

Lightolier has nation-wide distribution. There is at least one Lightolier dealer in your town, where you can examine adequate stocks of Lightoliers and also obtain Lightolier expert advisory service gratis.

If you don't know where the nearest Lightolier dealer is, write us and we will gladly tell you by return mail, and send you at the same time, our booklet, "The Secret of Entrancing Light".

For convenience of people in New York, our showrooms are accessibly located.

This modernized Georgian Corona Lightolier, 72 inches long, with 20-inch shade, is a model of correct proportions —finished in gold or silver plate, or Roman gold.

Lightolier
COMPANY, N.Y.
569 Broadway at Prince St.
LIGHTING FIXTURE and LAMP HEADQUARTERS

This is a Candle Lightolier in Colonial design. Length 36 inches, with a spread of 17 inches, and having five lights — finished in sterling silver plate.

*From the Farmer Lamp Collection.
Chinese antique bronze with shade
of hand woven old gold silk tapestry
and finial of carved red carnelian.*

IN transforming the poetically beautiful creations of Chinese masters into pieces which combine artistic virtu with usefulness, a true harmony of design is sustained with rare skill —the rich tones of precious Jade, Carnelian, Rose Quartz, Amethyst, Lapis, are delicately intensified in the exquisite settings which subtly express their form and spirit.

*From the Farmer Lamp collection
An Emerald Jade base exquisitely
carved in the best early Chinese
manner, complemented with shade.*

THE exquisite forms of ancient idols, stately vases and curious censers — the inimitably lovely masterpieces wrought by Chinese artists from precious Jade, Rose Quartz, Amethyst, Crystal, Torquoise, Carnelian,—are transformed with sympathetic feeling and comprehension into objects wherein the charm of rare and delicate beauty is enhanced by the value of usefulness.

BOUDOIR LAMPS
DRESDEN FIGURES

BASE — Glazed c h i n a figures, each figure decorated in colors with gold trim. Height figures 6 inches.

SHADE — Covered with rayon silk in the most delicate colors. Hand tailored, stretched, with lace design. Diameter 8½ inches.

	1 to 5 Prs, Per Pair	6 or more prs Ass't, Per Pair
No. E-1616B—Blue		
No. E-1616G—Green		

Twelve in carton; wt pr 3 lbs.

ASSORTMENTS
Per Ass't

No. E-1616A-6—Consists of 6 pairs, 3 pairs each of above lamps; wt ass't 18 lbs

create unrivaled beauty

These outstanding new Miller creations bring striking and unusual effectiveness

pure metal of the finest kind—rich bronzes, silvery pewters, shining brass (or, in a less expensive base with fine metal finishing)—each lamp, by the integrity of its design, becomes a triumph of artistry.

From the rich simplicity of Early American models, through the polished beauty of Georgian and Colonial re-creations, up to the startling smartness and novelty of the Futuristic which gives pace and architectural quality to modernistic rooms, the Miller Lamps of today interpret every important decorative period and grace every decorative mode. They run the gamut from stark Grecian form through the quaint relics of Victorian kerosene days.

Interior decorators value them for their fine craftsmanship and for their fidelity to period art.

Your Dealer is Displaying these Lamps
Be sure to see them

Ask for the display of Miller Lamps, by name, in visiting your favorite lamp department. All the lamps here shown are now on special exhibition in fine department stores and electric stores. Besides these illustrated you will find other attractive Miller models in various intriguing period designs, and you will be able to see for yourself the rich beauty of artistic modeling, the burnished loveliness of shining metal and the striking and individualistic effect which each of these lamps creates.

Should your dealer not carry Miller Lamps write us immediately for information and for prices.

THE MILLER COMPANY, Meriden, Conn.
"Pioneers in Good Lighting Since 1844"

A bridge lamp of pure polished Bronze or in alternating green brass, hand molded with a harmony of curve and line which Cellini himself might have envied. The rich simplicity of its form and the smartness of its arrow curve and shade of stretched taffeta make it an addition of distinction for any type of room. No. L248-5446.

Modernism incarnate, in color and design, this striking lamp whose bowl of oxblood trimmed with Flemish Bronze is complemented by a dome-shaped shade of Red silk Moire. The lamp is obtainable also in green trimmed with Oxidized French Pewter with shade of green Moire. No. L2909-5606.

The rare and perfect beauty of a Greek urn translated with unusual felicity into an outstandingly lovely table lamp. The exquisitely harmonious base is of colorful Spanish bronze. A rich shade of intriguing design developed in tan figured silk conveys an Oriental atmosphere and gives a distinctively exotic effect. No. L2910-5625.

The Greek urn again inspires the perfect symmetry of this breathlessly lovely base, richly molded of Oiled Brass and green. The elaborately simple shade, of outstanding balance and beauty, is of figured green-gold silk. The graceful ornament above provides an exquisite finishing touch. No. L2900-5628.

The Woofle Bird—a delightfully gay little conceit exceedingly popular for modernistic interiors, for Boudoirs and nurseries. In three amusing color-schemes: red body and tail, yellow trim, green eyes; green body, red and yellow tail, green eyes; yellow body, green and yellow tail, yellow and blue trim, red eyes. No. L2933.

OF DISTINCTIVE CRAFTSMANSHIP

An Early American candle-lamp (also charming in French Provincial interiors) developed with fine craftsmanship in Early Colonial brass. The shade is crackled parchment. No. L2746-5311.

A quaint electrified re-issue of the once-familiar Kerosene lamp, developed in four color-schemes to harmonize with any interior, makes a novel and acceptable Christmas gift. No. L2926.

A Georgian candle-lamp which will lend credit to any room of any period. It comes in silvery pewter and brass or in modernistic steel and brass. The shade is a charming soft silk. No. L2918-5623.

THE MILLER COMPANY, MERIDEN, CONNECTICUT

"Pioneers in Good Lighting Since 1844"

An intricate and charming example of distinctive craftsmanship. Candle lamp in antique Tudor gold with highly polished copper smoke bells. Hand finished with exquisite yet restrained decorative effect. No. L2862.

The outstanding new vogue for things Victorian is exemplified by this attractive old-copper re-creation of a Vestal Hand Kerosene lamp, to which the delightful printed parchment shade adds a finishing touch. No. L2928-5604.

Also a revival of things Victorian—the Vestal "parlor lamp" modeled in a lovely soft pewter and enlivened by the gala hunting scene which decorates the parchment shade. These lamps take a T-10 tubular bulb. No. L2927-5603.

A delightful Tudor two-branch candelabra, carved on exquisite proportions in Tudor Gold antique. The decorated parchment shade adds the last perfecting touch. No. 2911-5572.

Another delightful Georgian candle-lamp, authentically designed; most effective for desk, or console table. In silvery French pewter or in shining Flemish Bronze, with distinctive print parchment shade. No. L2914-5579.

MILLER *lamps*
OF DISTINCTIVE CRAFTSMANSHIP

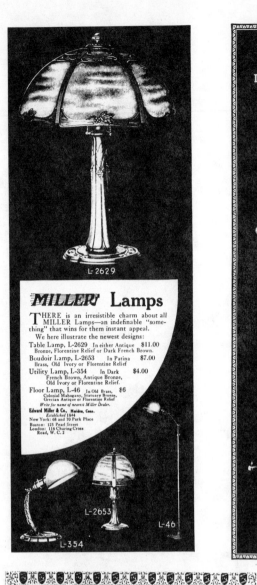

1245 55×26 cm
22 × 10"

Pendant S-975

Individuality in dining-room fixtures

THE unusual, coupled with practicability and durability, is achieved in this new type of dining-room fixture. The shades of a new material—Fabrikon—are hand decorated, permanent and imperishable; soft and colorful under illumination, and possessing a daylight value that adds greatly to the room furnishings. Side wall sconces to match the pendant either in polychrome gold or silver.

Small illustrations illustrate the side fixtures with double or single lights fitted with shades or shields as desired.

Tear-drop prisms both on the main and side fixtures add a pleasing color note. The polychrome gold bears the amber prism, and the silver fixture the turquoise.

Every Handel Lamp and fixture is stamped with the Handel name. Ask your lamp or fixture dealer or write us for further information.

THE HANDEL COMPANY, *MERIDEN, CONN.*

HANDEL *Lamps & Fixtures*

S-976   S-978   S-976

No. D3080.   ELECTRIC LAMP 1 Light

Height over all, 15½ inches.
Marble Base, Engraved Crystal Glass Fount.
Crystal Cut Glass Pillar, Crystal Pendants.
Crystal Cut Glass Frosted Shade.

» » VARIED LAMPS
by HANDEL *for* OLD
and NEW ROOMS » »

The lamp below is a modernistic and yet conservative design suitable for use with most types of furniture groupings. The base is of bright pewter in combination with black ebony, while the shade is hand-painted silver and black on skintex. Height 20" over all.

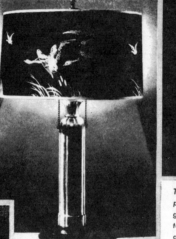

This Handel lamp was especially designed for use with furnishings of that period of our history usually referred to as the Federal era. The lamp, of exquisite workmanship, is styled with a hand-painted shade, covered and lined with honeydew silk. The base is finished in weathered old brass with eagles in Colonial gilt. Size 22¼" over all.

The Handel bridge lamp above correctly interprets the Empire period. The base is beautifully finished in empire green and antique gilt. It is 61" high and has an arm which raises and lowers and adjusts to any position. The shade is hand-painted under light gold pleated silk and lined with honeydew silk.

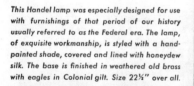

THE effectiveness of any given example of interior decoration is based primarily upon what it achieves in point of "suitability". This element of suitability has been taken into account in all Handel Lamp designs; some carry the fine things that tradition has brought them and yet strike a new note, while others are unrelated to tradition and are smartly new. All are keyed to our time and express our present feeling for elegance with restraint.

We cordially invite the architect, the decorator, and the home-owner to visit our showrooms or if that is not convenient, to write us a description of the type of lamp desired.

THE HANDEL COMPANY
MAKERS OF
HANDEL LAMPS & LIGHTING FIXTURES
200 Fifth Ave., New York • Meriden, Connecticut

No. 8901

BOUDOIR LAMP

Glass Vase on Cast Metal Mounting.

Colors of Vases: Rose, Blue, Black.
8" Parchment Shades in Ship or Flower Decoration.
Wired complete ....

No. 8909

BOUDOIR LAMP

Height over all, 14".
Width of Shade, 7".
Finishes: Ivory Polychrome with Blue, Rose or Gold Silk Inner Shade.
Wired complete ....

No. 8600

BOUDOIR LAMP

Pressed Flower Decorated Glass Shade.

Cast Metal Stand, Ivory or Gold.
Height over all, 14".
Wired complete ....

No. 8051

GLASS TORCHIERE

Height 10"   Width 4"
Decorated in Flame Tints.
Wired complete ....

No. 8020

PARROTT LAMP

Made entirely of Glass.
Colors: Red or Blue.
Wired complete ....
Put me on display and I'll speak for myself!

No. 8010

WISE OWL LAMP

Made entirely of glass.

Maple Vanity Lamp

These Vanity Lamps of rich, dark genuine maple are beautifully finished and reflect the Colonial spirit with the high-grade shade of parchment with silhouette decoration. These lamps are very effective used in pairs or separately. They are suitable anywhere for their soft, rich coloring blends with other furnishings. Stands 16 in. high; shade, 10 in. in diameter. Push-button socket. 5-ft. cord. Mlg. wt. 2 lbs.

2153 **$4** With *Purchase or Coupons*

Knight-in-Armor Lamp

All the romance of the days of chivalry. A beautiful Knight-in-Armor figure with an antique-parchment finish shade with coat of arms decoration. The Knight figure is 9 in. high, solid metal, finished in a beautiful satin silver-plate. The base is solid walnut. Shade is 10 in. in diameter. Silver-plated metal parts are non-tarnishable. Height over all, 15½ in. Mailing weight 6 lbs.

1612 **$7** With *Purchase or Coupons*

Figure Lamp

Two clownish, dancing figures form the amusing base for this decorative lamp of durable composition finished in verdi green. The 5-in. globe, white translucent glass; gives a soft glow of light. Complete with cord, plug and socket. The base is 8 in. wide and 6¾ in. deep; 12½ in. high over all. Mlg. wt. 12 lbs.

1707 **$5.50** With *Purchase or Cpns.*

Figure Lamp

4004 With **$4.50** *Purchase or Cpns.*

Strikingly beautiful, big lamp done in the modern manner that makes an excellent piece for radio, mantel or the console table. The two figures support a crackled, green glass globe. Made of hard, durable composition and finished in antique dull gold bronze. 19½ in. high. Base, 10 in. wide. Mailing weight 13 lbs.

Table Lamp

Decorative and useful Lamp for small tables or end tables. The shade is only 8 in. wide and 14 in. long so that this lamp will not look top heavy. Parchment with silhouette decoration. 2-candle style standard of heavy brass plate. About 14 in. high. Mlg. wt. 7 lbs.

2226 With **$6.50** *Purchase or Coupons*

Marvel Photo Lamp

This lamp is unique because it gives you a beautiful reproduction of any photograph or snapshot you send us, framed in a handsome, antique-gray metal frame complete with glass and lighting equipment ready for use when received.

A new process of photographing on "Translite" paper, hand-colored by trained artists in the natural coloring of hair, eyes, background etc. according to your instructions, results in a picture of greater depth, brilliancy and naturalness than has hitherto been possible. This reproduction is placed between glass with an electric bulb back of it. Your photo will be returned to you unharmed. Be sure to specify color of hair, eyes, etc. when sending photo. Any size photo is finished 8 in. high; 6 in. wide. Lamp is 10¼ in. high—antique gray finish. Completely wired. Bulb not furnished. Shipped from Kansas City, Mo. Mlg. wt. 4 lbs.

2400 With **$10** *Purchase or Coupons*

Illuminated Ship

The latest in ornaments for buffet or mantel. Beautiful all-metal ship in antique bronze, illuminated with 25-watt colored bulb in base which casts a soft glow through the port holes and upper parts of the ship. Also effective for console, radio, piano, etc. It will make a bright, attractive corner in any home. 12 in. high; 14 in. long over all. Complete with bulb.

Mailing weight 13 lbs.

3540 With **$10** *Purchase or Coupons*

FASCINATING-REALISTIC "SCENE-IN-ACTION" LAMPS

Artistic, fascinating, realistic! A view of Niagara Falls in natural colors. *When lighted the water rushes over the precipice in never-ending torrents* — the mist and constant churning of the water below the Falls is all shown with striking realism. Both American and Canadian Falls are shown. Concealed in the top is an incense cup. Operates on 105-120 volt current; complete with cord, plug and bulb. 9¼ in. high. Mlg. wt. 4½ lbs.

3451 With **$9.50** *Purchase or Coupons*

Other "Scene-in-Action" Lamps on page 211.

"Scene-in-Action" Aquarium

Distinguished from all other forms of animated lighting because the combination of fish swimming in the aquarium is in harmony with the "Scene-in-Action" in the base, which shows submarine life in natural state— fish swimming about, ocean currents briskly churning up a sandy bottom and every natural color brought out to best advantage. Metal base beautifully finished in dark bronze. Complete with cord and bulb ready to operate. Simply fill the aquarium with water and fish; you'll be delighted with the effect. 2-gallon aquarium. 14 in. high; 12 in. diameter. Mailing weight 20 lbs.

4357 With **$12** *Purchase or Coupons*

Scene-in-action is a new and original means of endowing beautiful pictures with life, color and movement—the latest contribution to artistic lighting. They brighten a dark spot, decorate the radio or mantel and are suitable for night lights, hall lights or wherever a soft light is desired. Nothing to get out of order. Fascinating Japanese scene of Fujiyama, the sacred mountain, and the enchanting, rippling lake below. In modern white metal frame, finished in antique silver. 13¾ in. wide; 9 in. high; 4¾ in. deep. Complete with bulb. Operates on 105-120 volts. Mailing weight 9 lbs.

1478 With **$12** *Purchase or Coupons*

Desk or Radio Lamp

Here is a handy lamp with many uses—for the radio or desk, as a reading lamp, or, with the shade turned up providing indirect lighting, it makes a good light when playing cards. It is highly ornamental, well made of metal in bronze finish. Even the shade is metal. Push-button socket. Height, 12 in.

Mlg. wt. 5 lbs. **4940** With **$3** *Purchase or Coupons*

De Luxe Desk Lamp

This Desk Lamp is the aristocrat of its type and is a worth-while addition to any desk or table. The shade is adjustable to all angles by means of a ball swivel. The neatly designed base is grooved for pen and pencil. Finished in antique-green. Height, 14 in.; base, 6¼ x 5 in. Shade is 8½ in. wide, equipped with chain-pull socket. 7-ft. cord with easy-pull Bakelite cap.

Mailing weight 4½ lbs.

1655 With **$6** *Purchase or Coupons*

Electric Shadow Lamps

No. 6711 SHADOW LAMP. Stand 12½ in. high, base 4½x5 in. of fancy design. Made of cast metal in Mahogany Brown with Patina Green shading. Art lady figure on the stand, with ornamented frosted glass. Electrical equipment in back.

Each ....

No. 6709 SHADOW LAMP. 9½ in. high. Base 8x2 in. Made of one piece cast metal, finished in Mahogany Brown. Fancy scroll open work design with figure of tiger at one end. Frosted glass with electrical equipment in back.

Doz ....

Each ....

No. 5583 SHADOW LAMP. 11½ in. high, in weather brown, with green patina trim. Figure is full fashioned, and stands apart from the glass, felt bottom.

Per doz. ....

Each ....

No. 6710 SHADOW LAMP. 8½ in. high, base 8x4 in., representing a rock with the figure of a Panther. Cast of metal in Mahogany Brown finish, and has oval frosted glass, ornamented with tropical trees. Electrical equipment in back.
Each ....

No. 5585 SHADOW LAMP. Base and frame of cast metal with artistic outlined figures within the frame, which is fitted with opalescent glass in frosted effect. Electrical equipment in the back, complete with cord. 8½ in. high, in old gold with green trim
Per doz. ....

Each ....

No. 5277 TABLE LAMP. Complete with shade. Fancy cast base and novelty frame. Finished in antique gold plate. **Two-light Candle** model fixture, with Vidrio Onyx ornament. Tex Parchment shade, with laced borders ornamented with large colored picture, mounted on wire frame, securely fastened with fancy screw top. Standard electric equipment with cord. Packed **singly** or **six** in a case.

No. 6708 SHADOW LAMP. 8½ in. high. Base 8¾x2 in. Made of cast metal in Mahogany Brown finish with Patina Green shading. Design of Overland Coach, has frosted glass with electrical equipment in back.

No. 6712 AQUARIUM. Base 8¾ x4½ in. Cast of metal with an Art Maiden figure at each end, and crystal fish bowl set in center, which is removable. Finished in Mahogany Brown.

No. 3249 Faries Verdelite Portables
With Adjustable and Detachable Green Glass Shade

Swing portable, vertical adjustment with adjustable and detachable green glass shade. For 25 to 60-watt lamp. Extends 10 inches. Wired with 9 feet of cord, pull chain socket and plug. Standard package, 12 assorted.

Price, No. 3249 Brushed Brass............each
Price, No. 3249 Statuary Bronze............each

No. 3230 Faries Verdelite Portables

With Adjustable and Detachable Green Glass Shade

Fancy solid brass portable. For flat top desk or table. Height, 15 inches to center of shade. For 25 to 60-watt lamp. Wired with nine feet of silk cord, pull chain socket and plug. Standard package, 1 dozen assorted.

Cat No.	Description	Price Each
3230	Brushed Brass with Black Relief....	
3230	Statuary Bronze, Green Relief....	

No. 3245 Faries Verdelite Portables

With Adjustable and Detachable Green Glass Shades

Plain solid brass portable. Height, 16½ inches to center of shade. For 25 to 60-watt lamps.

Wired with nine feet of silk cord, pull chain sockets and plug.

Std. pkg., 12 assorted.

Cat No.	Description	Price Each
3245	Brushed Brass	
3245	Statuary Bronze.....	

No. 3247 Faries Verdelite Portables

With Adjustable and Detachable Green Glass Shades

Fancy cast brass portable. Height, 15½ inches to center of shades. For 25 to 60-watt lamps. Wired with 9-foot silk cord, pull chain sockets and plug. Standard package, 12 verdelites assorted.

No. 3247, Brushed Brass, Black Relief..each
No. 3247, Statuary Bronze, Green Relief.each
No. 3247, Statuary Bronze and Gold.each

No. 100 Pittsburgh Permaflectors

For windows having a depth (plate glass to background) ranging from ½ to ¾ the distance from floor to reflector. Light flux covers from 0 to 90°.

Strong concentration of light between 5 and 25° gives high floor illumination near plate glass.

Ideal for windows in which trim is carried high; also works well in deeper windows.

Size of lamp, 150-200-watt, clear. Form O, 2¼-inch holder.

Height, 8⅜ inches; width, 9⅛ inches; front to back, 8½ inches; center to back, 3⅝ inches.

Carton, 10. Standard package, 40; weight, 87 pounds.

No. 100, Reflector and Adapter No. 1234........each

No. 51 Pittsburgh Permaflectors

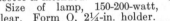

For high shallow windows, medium high trim, island windows or windows with upper part of background of glass. Concentrating; light cuts off sharply on 55° line. Exceptional concentration of light in 5 to 25° angle, insures effective illumination on floor of high and shallow windows for which reflector is intended.

Size of lamp, 150-200-watt, clear. Form O, 2¼-in. holder.

Height, 8¼ inches; width, 9⅜ inches; front to back, 9⅝ inches; center to back, 3⅞ inches.

Carton, 8. Standard package, 32; weight, 83 pounds.

No. 51, Reflector and Adapter No. 1234........each

No. 52 Pittsburgh Permaflectors

For high shallow windows, medium high trim, island windows or windows with upper part of background of glass. Designed especially for inserting in ceiling; bottom of reflector is level and can be installed flush with ceiling.

Size of lamp, 150–200 watt, clear.

Form O, 2¼-inch holder.

Height, 7¹¹⁄₁₆ inches; width, 9 inches; front to back, 9½ inches; center to back, 3½ inches.

Carton, 8. Standard package, 32; weight, 82 pounds.

No. 52, Reflector and Adapter No. 1234........each

No. 55 Pittsburgh Permaflectors

For medium size shallow windows, medium high trim, island windows or windows with upper part of background of glass. Concentrating; light cuts off between 65 and 70°.

Exceptional concentration in the 5 to 35° angle, insuring effective illumination on floor of shallow window for which reflector is intended.

Lamps, 100-watt, A23 (inside frosted) and 150-watt, clear. Form O, 2¼-inch holder.

Height: with adapter in 150-watt position, 6⅞ inches; with adapter in 100-watt position, 6³⁄₁₆ inches; diameter opening circular, 8½ inches.

Carton, 10. Standard package, 40; weight, 76 pounds.

No. 55, Reflector and Adapter No. 1256........each

Faries Adjustable Portables

The Standard for Quality

All Faries portables are equipped with 9 feet approved cord, standard socket and rubber plug that will not break. Passed by Underwriters. Bases are heavily felted.

No. 789

A popular strong desk lamp with one piece, heavy brass shade that is adjustable and full size, 8¾ inches long. A small lamp that gives adequate light. Height 14 inches. Standard package 6.

Bronze ....
Mahogany and Green..

No. 1012

A utility lamp with heavy brass covered base. Adjustable to any position. Has large tubing and strong one piece joints. Maximum height 25 inches, extends 24 inches. Has patented shade with slip-on holder that swivels on socket. Key socket. Standard Package 6.

Brushed Brass ....
Statuary Bronze ....

No. 3263 Telescopic

High quality telescoping floor portable with flexible arm for Professional Men and Draftsmen.

Stem in pedestal swings around horizontally or slides up and down, holding any position without set screws. Large size, patented metal shade (No. 32) swivels on socket. Adjustable in height 2½ to 5 feet. Has 9-inch brass flexible arm and heavy base, felted. Standard package 6.

Brushed Brass ....
Statuary Bronze...........................
White Enamel and Chrome.........

No. 1571

New modern design. Very neat but strong—made of brass. Large round shade 7 inches in diameter, swivels on ball joint. Heavy weighted base, diameter 5 inches. Height 14 inches. Push socket. Standard Package 6.

All Polished Chrome....................
Polished Chrome with Satin Copper Shade ....
Polished Chrome with Cloister Bronze Shade ....

No. 3266 Non-Telescopic

The ideal floor portable with non-telescoping stem and flexible arm, for students and home use.

Stem is stationary, but 9-inch brass flexible arm permits adjustability in height and extension. Large size, patented metal shade (No. 32) swivels on socket to focus light where wanted. Height, 48 inches. Has heavy base, felted. Standard package 6.

Brushed Brass ....
Statuary Bronze ....

HERE ARE SOME BIG LEADERS FOR YOU IN ELECTRIC LAMPS
Complete With An Assortment of Shades In Colors Blue, Rose, Taupe, Mulberry or Beaver

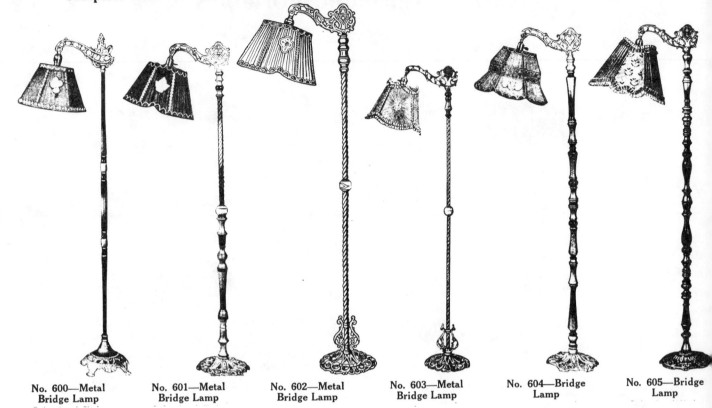

No. 600—Metal
Bridge Lamp

No. 601—Metal
Bridge Lamp

No. 602—Metal
Bridge Lamp

No. 603—Metal
Bridge Lamp

No. 604—Bridge
Lamp

No. 605—Bridge
Lamp

ANOTHER PAGE OF BIG LEADERS IN BRIDGE LAMPS COMPLETE
Assorted Colors of Shades—Blue, Rose, Taupe, Mulberry, Etc.

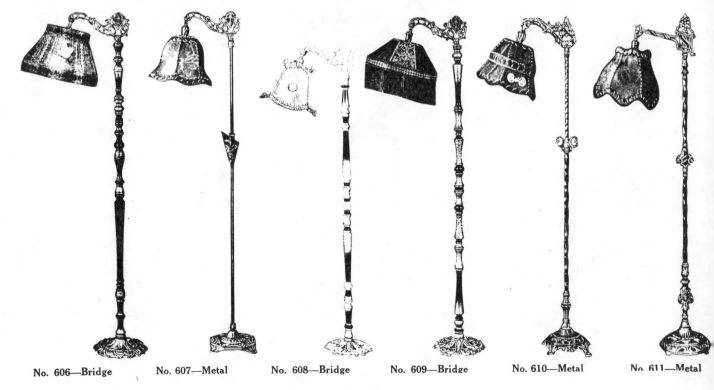

No. 606—Bridge

No. 607—Metal

No. 608—Bridge

No. 609—Bridge

No. 610—Metal

No. 611—Metal

BEAUTIFULLY TAILORED FANCY SHADES IN BEAVER, ROSE, BLUE, OR TAUPE ON ARTISTICALLY MADE JUNIOR BASES, COMPLETE ON METAL OR WOOD. POLYCHROME FINISH.

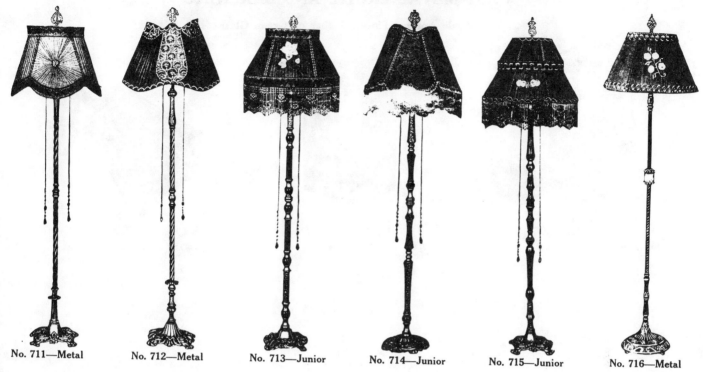

No. 711—Metal No. 712—Metal No. 713—Junior No. 714—Junior No. 715—Junior No. 716—Metal

VERY LOW PRICED LEADERS IN JUNIOR LAMPS COMPLETE

With Beautiful Assortment of Fancy Shades In Latest Colors in Taupe, Beaver, Blue, or Mulberry

Page 92 explains how you can increase your profits without any extra expense.

No. 700—Metal Junior Lamp No. 700½—Junior Lamp No. 703—Junior Lamp No. 704—Metal Junior Lamp

CAST METAL LIGHTS AND BRACKETS

Verde Antique Finish, Amber or Green Glass

BRACKET

No. 2000 5 x 9 in., Complete with Fixtures ....Each,

PORCH LIGHT

No. 5000 8 x 10 in., Complete with Fixtures ....Each,

BRACKET

No. 2005 5½ x 7 in., Complete with Fixtures ....Each,

HALL LIGHTS

No. 5010 4½ x 9 in., Complete with Fixtures ....Each,

No. 5005 Diameter 12 in., Height 14 in., Complete with Fixtures, Each,

No. 5015 Diameter 9 in., Length 30 in., Complete with Fixtures, Each,

No. 5187

Height 5 Inches; Base 6 Inches; Glass 6 x 3¼ Inches; Holder 4 Inches.

	Bare	Complete	Glass
No. 5187...........Each,			

No. 5911

Height	....	14 Inches
Base	....	7 Inches
Casing	....	2 Inches
Glassware	....	8 x 4 Ball
Holder	....	4 Inches

	Bare	Complete	Glass
No. 5911 ....Each,			

Nos. 74014 to 74020

Length 48 Inches, Body 8 Inches, Spread 32 Inches

	Bare	Complete	Glass
No. 74014	Four Light ...Each,		
No. 74016	Six Light ....Each,		
No. 74018	Eight Light ...Each,		
No. 74020	Ten Light ....Each,		

No. 5912

Height	....	16 Inches
Base	....	6 Inches
Casing	....	2½ Inches
Glassware	....	8 x 4 Ball
Holder	....	3¼ Inches

	Bare	Complete	Glass
No. 5912 ....Each,			

Above Prices are for Gilt, Brush Brass, or Oxidized Brass. Oxidized Copper Add 10%. Special Finishes at Special Prices. Complete on Above Means, "Wired Complete with Sockets." *Glassware, Holders,* and *Electric Bulbs* are Extra.

COMBINATION BRACKETS

No. 2200
Extends 12 Inches, Spread 16 Inches.

Bare   Complete   Glass

No. 2200   Two Gas, Two Electric.....Each,
Fitted for 3¼ Inch Electric and 4 Inch Gas Glass.

No. 2205
Extends 12 Inches, Spread 15 Inches.

Bare   Complete   Glass

No. 2205   One Gas, Two Electric.....Each,
Fitted for 2¼ Inch Electric, and 4 Inch Gas Glass.

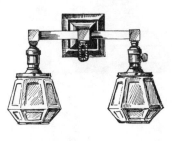

No. 2210
Extends 7 Inches, Spread 9 Inches.

Bare Complete Glass

No. 2210   One Gas, One
Electric. Each,
Fitted for 3¼ Inch Glassware.

No. 2215
Extends 6 Inches, Spread 9 Inches

Bare   Complete   Glass

No. 2215   One Gas, Two
Electric. Each,
Fitted for 2¼ Inch Electric and 4 Inch Gas Glass.

No. 2220
Extends 8 Inches, Spread 12 Inches.

Bare Complete Glass

No. 2220   One Gas, Two
Electric. Each,
Cast Iron Flame Complete
Extra ....
Fitted for 2¼ Inch Glass.

No. 2225
Extends 12 Inches, Spread 15 Inches

Bare   Complete   Glass

No. 2225   Two Gas, Two Electric.....Each,

Fitted for 2¼ Inch Electric, and 4 Inch Gas Glass.

No. 2230
Extends 6 Inches, Spread 9 Inches.

Bare   Complete   Glass

No. 2230   Two Gas, Two Electric.....Each,
Candles Extra ....Each,
Fitted for 2¼ Inch Gas Glass.

Above Prices are for Gilt, Brush Brass, or Oxidized Brass. Oxidized Copper, Add 10%. Special Finishes at Special Prices. Complete on Above Means, "Wired Complete With Sockets." *Glassware, Holders,* and *Electric Bulbs* are extra.

No. 2055

Extends 5 Inches

e ....Each,
nplete ....Each,
ss ....

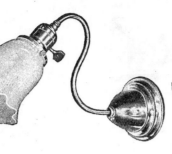

No. 2060

Extends 7 Inches

Bare ....Each,
Complete ....Each,
Glass ....

No. 2065

Extends 8 Inches

Bare ....Each,
Complete ....Each,
Glass ....

No. 2070

Extends 7 Inches

Bare ....Each,
Complete ....Each,
Glass ....

No. 2075

Extends 7 Inches

e ....Each,
nplete ....Each,
ss ....

No. 2080

Extends 6 Inches

Bare ....Each,
Complete ....Each,
Glass ....

No. 2085

Extends 6 Inches

Bare ....Each,
Complete ....Eaeh,
Glass ....

No. 2090

Extends 6 Inches

Bare ....Each,
Complete ....Each,
Glass ....

No. 2095

Extends 6 Inches
Spread 9 Inches
Two Light

e ....Each,
nplete ....Each,
ss ....

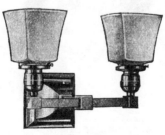

No. 2100

Extends 7 Inches
Spread 9 Inches
Two Light

Bare ....Each,
Complete ....Each,
Glass ....

No. 2105

Extends 4½ Inches
Spread 7 Inches
Two Light

Bare ....Each,
Complete ....Each,
Glass ....

Above Prices are for Gilt, Brush Brass, or Oxidized Brass. Oxidized Copper, Add 10%. Special Finishes at Special Prices. Complete on
ve Means, "Wired Complete With Sockets." *Glassware, Holders,* and *Electric Bulbs* are extra.

ELECTRIC BRACKETS

No. 2010

Extends 9 Inches

Bare ....Each,
Complete ....Each,
Glass ....

No. 2015

Extends 9 Inches

Bare ....Each,
Complete ....Each,
Glass ....

No. 2020

Extends 7 Inches
Holder 4 Inches

Bare ....Each,
Complete ....Each,
Glass ....

No. 2025

Extends 6 Inches
Spread 10 Inches

Bare ....Each,
Complete ....Each,
Glass ....

No. 2030

Extends 7 Inches
Spread 10 Inches

Bare ....Each,
Complete ....Each,
Glass ....

No. 2035

Extends 9 Inches
Spread 9 Inches

Bare ....Each,
Complete ....Each,
Glass ....

No. 2040

Extends 8 Inches
Spread 12 Inches

Bare ....Each,
Complete ....Each,
Glass ....

No. 2045

Extends 8 Inches
Spread 4¾ Inches

Bare ....Each,
Complete ....Each,
Electric Lamps ....

No. 2050

Extends 5 Inches
Spread 5½ Inches

Bare ....Each,
Complete ....Each,
Prism Glass ....

Above Prices are for Gilt, Brush Brass, or Oxidized Brass. Oxidized Copper, Add 10%. Special Finishes at Special Prices. Complete on Above Means, "Wired Complete With Sockets." *Glassware, Holders,* and *Electric Bulbs* are extra.

MAZDA FIXTURES

No. 1523

Length, 15 Inches. Spread, 14 Inches.

		Bare	Complete	Glass
No. 1522	Two Light ....Each,			
No. 1523	Three Light ....Each,			

No. 1294

Length, 16 Inches. Spread, 15 Inches.

		Bare	Complete	Glass
No. 1292	Two Light ....Each,			
No. 1294	Four Light ....Each,			

No. 1533

Length, 15 Inches. Spread, 16 Inches.

		Bare	Complete	Glass
No. 1532	Two Lights ....Each,			
No. 1533	Three Lights ....Each,			

No. 1084

Length, 15 Inches. Spread, 18 Inches.

		Bare	Complete	Glass
No. 1082	Two Lights ....Each,			
No. 1084	Four Lights ....Each,			

Above Prices are for Gilt, Brush Brass, or Oxidized Brass. Oxidized Copper Add 10%. Special Finishes at Special Prices. Complete on Above Means, "Wired Complete with Sockets." *Glassware, Holders,* and *Electric Bulbs are Extra.* Fixtures are Always Shipped Bare Unless Otherwise Ordered.

COMBINATION PENDANTS

No. 3200
Length 33 Inches

No. **3200** One Gas, One Electric.....Each,
No. **3205** One Gas, One Electric.....Each,

No. 3205
Length 33 Inches
Bare   Complete   Glass

No. 3210
Length 33 Inches

No. **3210** One Gas, One Electric.......Each,
No. **3215** One Gas, One Electric.......Each,

No. 3215
Length 33 Ins., Spread 10 Ins.
Bare   Complete   Glass

Length 33 Inches, Spread 12 Inches
No. **3225** One Gas, One   Bare Complete Glass
Electric...Each,

Length 33 Inches, Spread 10 Inches
No.**3230** One Gas One   Bare Complete Glass
Electric.Each,

Above Prices are for Gilt, Brush Brass, or Oxidized Brass. Oxidized Copper, Add 10%. Special Finishes at Special Prices. Complete **on** Above Means, "Wired Complete With Sockets." *Glassware, Holders,* and *Electric Bulbs* are extra

ELECTRIC DINING ROOM FIXTURES

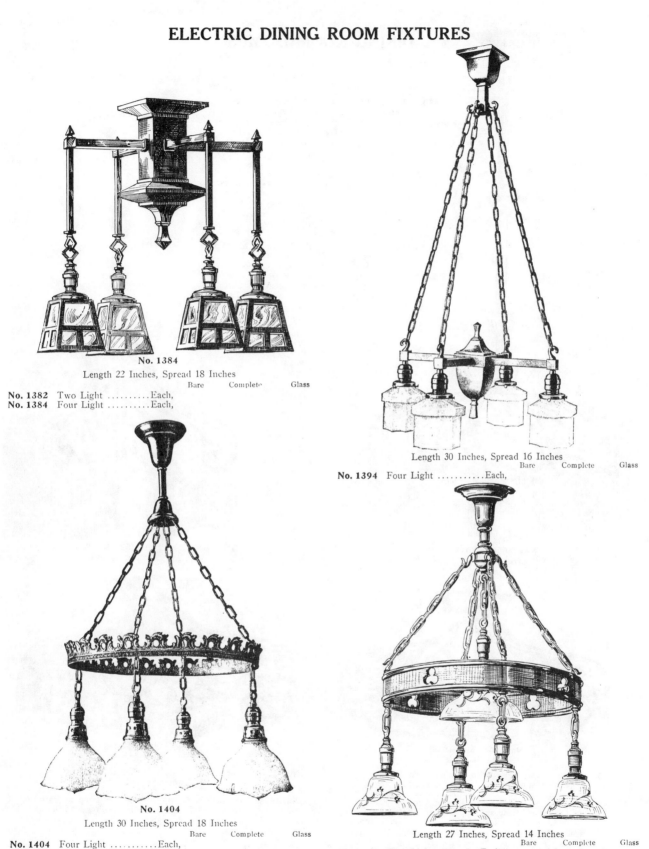

No. 1384

Length 22 Inches, Spread 18 Inches

		Bare	Complete	Glass
No. 1382	Two Light ....Each,			
No. 1384	Four Light ....Each,			

Length 30 Inches, Spread 16 Inches

		Bare	Complete	Glass
No. 1394	Four Light ....Each,			

No. 1404

Length 30 Inches, Spread 18 Inches

		Bare	Complete	Glass
No. 1404	Four Light ....Each,			
No. 1405	Five Light ....Each,			

Length 27 Inches, Spread 14 Inches

		Bare	Complete	Glass
No. 1415	Five Light ....Each,			

Above Prices are for Gilt, Brush Brass, or Oxidized Brass. Oxidized Copper Add 10%. Special Finishes at Special Prices. Complete on Above Means, "Wired Complete with Sockets." *Glassware, Holders,* and *Electric Bulbs* are Extra.

DINING ROOM DOMES

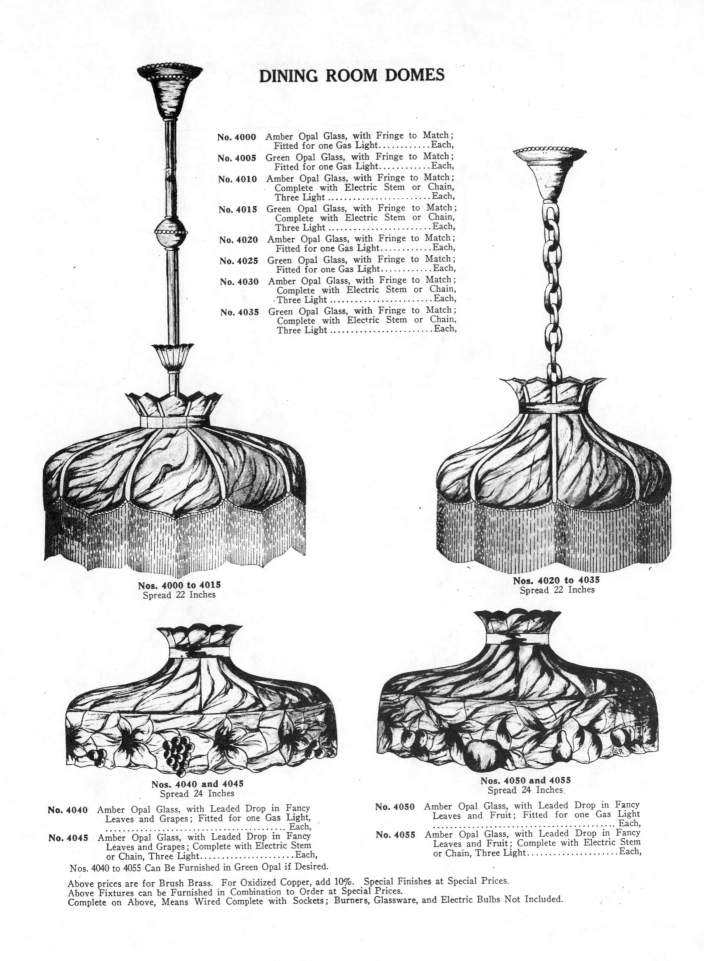

No. 4000   Amber Opal Glass, with Fringe to Match;
Fitted for one Gas Light...........Each,

No. 4005   Green Opal Glass, with Fringe to Match;
Fitted for one Gas Light...........Each,

No. 4010   Amber Opal Glass, with Fringe to Match;
Complete with Electric Stem or Chain,
Three Light ....Each,

No. 4015   Green Opal Glass, with Fringe to Match;
Complete with Electric Stem or Chain,
Three Light ....Each,

No. 4020   Amber Opal Glass, with Fringe to Match;
Fitted for one Gas Light...........Each,

No. 4025   Green Opal Glass, with Fringe to Match;
Fitted for one Gas Light...........Each,

No. 4030   Amber Opal Glass, with Fringe to Match;
Complete with Electric Stem or Chain,
Three Light ....Each,

No. 4035   Green Opal Glass, with Fringe to Match;
Complete with Electric Stem or Chain,
Three Light ....Each,

Nos. 4000 to 4015
Spread 22 Inches

Nos. 4020 to 4035
Spread 22 Inches

Nos. 4040 and 4045
Spread 24 Inches

Nos. 4050 and 4055
Spread 24 Inches

No. 4040   Amber Opal Glass, with Leaded Drop in Fancy
Leaves and Grapes; Fitted for one Gas Light,
....Each,

No. 4045   Amber Opal Glass, with Leaded Drop in Fancy
Leaves and Grapes; Complete with Electric Stem
or Chain, Three Light....................Each,

Nos. 4040 to 4055 Can Be Furnished in Green Opal if Desired.

No. 4050   Amber Opal Glass, with Leaded Drop in Fancy
Leaves and Fruit; Fitted for one Gas Light
....Each,

No. 4055   Amber Opal Glass, with Leaded Drop in Fancy
Leaves and Fruit; Complete with Electric Stem
or Chain, Three Light....................Each,

Above prices are for Brush Brass. For Oxidized Copper, add 10%. Special Finishes at Special Prices.
Above Fixtures can be Furnished in Combination to Order at Special Prices.
Complete on Above, Means Wired Complete with Sockets; Burners, Glassware, and Electric Bulbs Not Included.

Length 54 Inches, Spread 24 Inches, Depth of Dome 11 Inches.

Complete

No. 4065 Amber, with Canary Trimmings..............Each,

Length 54 Inches, Spread 24 Inches, Depth of Dome 11 Inches.

Complete

No. 4070 Amber, with Ruby Green Strip..............Each,

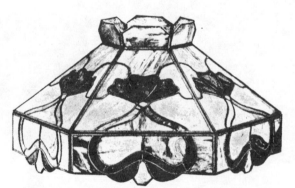

Length 54 Inches, Spread 24 Inches, Depth of Dome 13 Inches.

Complete

No. 4075 Amber Back Ground, Ruby Green Flowers...Each,

Length 54 Inches, Spread 24 Inches, Depth of Dome 13 Inches.

Complete

No. 4080 Amber and Green Background, Purple and Ruby
Grapes ....Each,

Domes can be Furnished with Different Color Art Glass to Order. Also can be Furnished with Closed or Open Tops to be Used for Gas **or** Electric.

Special Lengths can be Furnished to Order. 54 Inch Length and Brush Brass Finish Will Always be Sent, Unless Otherwise Specified.

54 Inches is Standard Length for 9 Foot Ceiling.

Above Prices are for Brush Brass, or Oxidized Brass. Oxidized Copper, add 10%. Special Finishes at Special Prices. Complete on Above Means Wired Complete with Sockets. *Electric Bulbs* are extra.

CRYSTAL GIFT LAMPS

AL7950 Crystal Vanity Lamps, Per Pair
Three crystal balls on mirror base. Pleated eggshell clair de lune shade, trimmed with moiré ribbon. 8 inches in diameter. Height 13 inches. **Sold only as a pair.** A splendid match for **crystal table lamp** illustrated at right.

AL7951 Crystal Colonial Lamp ....
Crystal body, handle of polished brass, and crystal colonial chimney. Attractive yellow shade to match the polished brass. Shade 12 inches. Height 17 inches.

AL7952 Crystal Table Lamp ....
Three crystal balls on mirror base. Modern table lamp matching the vanity lamp illustrated at left. Eggshell clair de lune shade, trimmed with moiré ribbon, 10 inches in diameter. Height 16 inches.

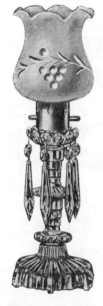

AL7953 Crystal Vanity Lamp, Per Pair
Diamond cut vanity lamp with claire de lune shade, ruching trimmed, eggshell color, 8 inches in diameter. Height 15½ inches. Sold only as a pair.

AL7954 Crystal Colonial Lamp ......Per Pair
Waterford design cut crystal. Early American glass chimney with cut prisms. Height 13 inches. Sold only in pairs.

AL7955 Crystal Colonial Lamp ......Per Pair
Early American design, glass chimney with cut prisms. Height 14 inches. Sold only in pairs.

AL7956 Cystal Vanity Lamp, Per Pair
Hobnail design glass vanity lamp with dotted Swiss drum shade over parchment. Eggshell color shade, 8 inches in diameter. Height 14 inches. Sold only as a pair.

PROCESSED ITALIAN WHITE MARBLE LAMPS

*Something
New and
Attractive*

—

*Shows Up
Beautifully
When
Lighted*

—

*Very Low
in Price*

No. 299E—Six Table Lamp Assortment as shown

Processed Marble Boudoir Lamps In Assorted Designs.   Height 17 Inches Over All.   Marble Is Tinted in Browntone. Wired with Push-Button Sockets, Cord and Separable Plug.   Hand Painted Parchment Glasse Shades. This Assortment Is a Real Value, Giving You Lamps at Extremely Low Prices. They Should Prove Big Profit Makers and Sell in Big Quantities.

No. 836 Piano Lamp; Green Bronzed Adjustable Swing Stand, with Lily Shade; Wired Complete....Each,

No. 1068 Piano or Desk Lamp, Polished Brass; Wired Complete ....Each,

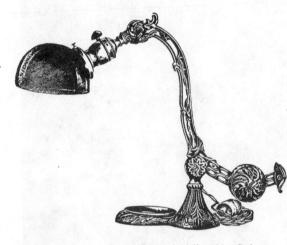

No. 936 Piano Lamp; Green Bronzed Adjustable Swing Stand; Green Cased Half Round Shade; Wired Complete ....Each,

No. 632 Jack Frost Junior Portable Lamp; Height 13 Inches; Shade 7½ Inches; Wired Complete; with Electric Bulbs ....Each,

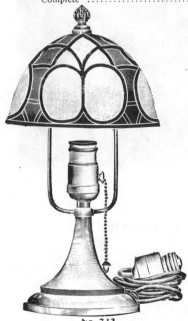

No. 732

Portable Lamp; Brush Brass; With Chain Pull Socket; Wired Complete....Each,

No. 832

Portable Lamp; Brush Brass; with Chain Pull Socket; Wired Complete....Each,

No. 932

Same as No. 832, but with Plain Globe ....Each,

No. 732

No. 832

Above Lamps are Wired Complete, and Include Sockets, Holders and Glassware. Electric Bulbs are Extra Except Where Noted.

No. 2002 Wall Luminarie, twin lights, assorted glazes.

No. 2003 Wall Luminarie, single light, assorted glazes.

No. 2004 Wall Luminarie, Gargoyle, all matte glazes.

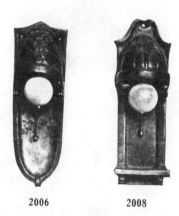

2006   2008

No. 2006 Renaissance Wall Light, verte antique splash, height 13 inches, complete with socket.
No. 2008 Colonial Wall Light, café au lait, height, 13 inches, complete with socket.

No. 19 Two Light Lamp, height 19 inches, leopard skin crystal.

No. 21 Two Light Lamp, height 18 inches, shade diameter 13½ inches, green flambé.

Mustard matte and black flambé, height 16½ inches.

No. 203 One Light Lamp, verte antique, height 16 inches.

ELECTRIC PORTABLES
With Art Glass Shades

No. 1311
Height 18 Inches, Shade 14 Inches.
Complete with Key Socket and Six Feet of Cord; Without Electric Bulb.
No. 1311G Brush Brass, Green Shade ....Each,
No. 1311A Brush Brass, Amber Shade ....Each,

No. 1300
Height 24 Inches, Shade 12 Inches.
Complete with Key Socket and Six Feet of Cord; Without Electric Bulb.
No. 1300G Brush Brass, Green Shade...............Each,
No. 1300A Brush Brass, Amber Shade...............Each,
No. 1312G Brush Brass, Green Shade...............Each,
No. 1312A Brush Brass, Amber Shade...............Each,
No. 1312 is Same as No. 1300 with Plain Border on Shade.

No. 1309
Height 24 Inches, Shade 12 Inches.
Side Brass Ornaments.
Complete with Key Socket and Six Feet of Cord; Without Electric Bulb.
No. 1309G Brush Brass, Green Shade...............Each,
No. 1309A Brush Brass, Amber Shade...............Each,

No. 1315
Height 24 Inches, Shade 12 Inches.
Side Ornaments.
Complete with Key Socket and Six Feet of Cord; Without Electric Bulb.
No. 1300G Brush Brass, Green Shade...............Each,
No. 1300A Brush Brass, Amber Shade...............Each,

THREE BEAUTIFUL TABLE LAMPS

No. 636—Table Lamp

Complete With Shade. Lamp is 28 Inches High.
Gold Polychrome Finish with Enamel Shading.
Stippled and Striped in Black and Gold.
Wired Complete with Two Light Pull Chain Fixture,
Cord and Separable Plug.
Shade is 19x14 Inches. Semi-Pleated Silk Georgette
Top with Embroidered Panels.
Charmeuse Lining. Wide Braid Borders. Heavy
5-Inch Silk Fringe Over Picot Valance.
Colors: Blue Over Rose and Rose Over Gold.

Factory No. 8550. Each......................

No. 637—Table Lamp
Complete With Shade.
Lamp Is 28 Inches High. Has Wide Base.
Is Plated and Hand Buffed In Gold.
Wired with Two Light Pull Chain Cluster,
Cord and Separable Plug.
Shade Is 15x15 Inches.
Scalloped Shape. Embossed Changeable
Two-Tone Rayon Top. Drum Lined.
Silk and Tinsel Braid Trimming.
Colors: Rose Green Over Rose and Gold
Over Rose.
Factory No. 9640. Each..........

No. 638—Table Lamp
Complete With Beautiful Hand Painted
Glasse Shade.
Height 28 Inches.
Plated Stand of Vidrio Onyx and Har-
monizing Metal Parts In Gold Color.
Wired Complete with Two Light Pull
Chain Fixture, Cord and Separable Plug.
Shade Is 19x13 Inches.
Hand Painted Glasse Covered with Small
Beads to Give an Iridescent Lighting
Effect. 5 Inch Beaded Fringe.
Factory No. 4925. Each.........

PRICE GUIDE

PAGE 1
E500 - $400 +

PAGE 2
E1070 - $400 +
E1080 - $400 +
E1100 - $500 +
E470-680 - $500 +

PAGE 3
E910 - $500 +
E460 - $500 +
E470 - $400 +
E490 - $400 +

PAGE 4
ALL $400 +

PAGE 5
ALL $400 +

PAGE 6
E960 - $400 +
E1060 - $400 +
E570 - $300 +
E620 - $300 +

PAGE 7
ALL $400 +

PAGE 8
E800 - $300 +
E120 - $300 +
E540 - $400 +
E580 - $400 +

PAGE 9
ALL $400 +

PAGE 10
ALL $400 +

PAGE 11
E1020 - $400 +
E1030 - $400 +
E900 - $500 +
E790 - $400 +

PAGE 12
E630 - $1,000 +
E710 - $400 +
E680 - $1,200 +
E970 - $500 +

PAGE 13
ALL $400 +

PAGE 14
ALL $500 +

PAGE 15
ALL $400 +

PAGE 16
E180 - $400 +
E60 - $400 +
E321 - $300 +
E280 - $400 +

PAGE 17
ALL $400 +

PAGE 18
ALL $400 +

PAGE 19
ALL $400 +

PAGE 20
ALL $400 +

PAGE 21
E10 - $300 +
E20 - $400 +
E5 - $200 +
E380 - $300 +
E345B - $30 +
E370B - $30 +
#3750 - $50 +
#3724 - $30 +
E335B - $20 +
E340B - $30 +
#11021 - $40 +

PAGE 22
E310 - $600 +
E320 - $300 +
E340 - $200 +
E370 - $200 +
E351 - $250 +

PAGE 23
E410 - $150 +
E330 - $200 +
E390 - $200 +
E402 - $150 +
E401 - $200 +
E420 - $150 +
E400 - $200 +
E350 - $200 +

PAGE 24
E891 - $100 +
E840 - $100 +
E890 - $100 +
E301 - $50 +
E860 - $50 +
#1174 - $30 +
#1012 - $25 +
E406 - $50 +
#2704 - $30 +
E300 - $50 +

PAGE 25
S6 - 162 - $50 +
S6 - 163 - $50 +
S6 - 164 - $75 +
S6 - 165 - $100 +
S6 - 166 - $75 +

PAGE 26
ALL $50 +

PAGE 27
S6 - 323 - $50 +
ALL OTHERS $35 +

PAGE 28
S6 - 389 - $35 +
S6 - 390 - $125 +
S6 - 391 - $35 +
S6 - 392 - $75 +

PAGE 29
S6 - 107 - $350 +
S6 - 108 - $400 +

PAGE 30
S6 - 375 - $400 +
S6 - 376 - $400 +
S6 - 377 - $100 +
S6 - 378 - $400 +
S6 - 379 - $400 +

PAGE 31
ALL $100 +

PAGE 32
S6 - 109 - $350 +
S6 - 110 - $500 +

PAGE 33
A - 106 - $75 +
A - 107 - $75 +
A - 108 - $100 +
A - 109 - $25 +
A - 110 - $40 +

PAGE 34
A - 112 - $75 +
A - 113 - $125 +
A - 115 - $125 +
A - 118 - $100 +

PAGE 35
A - 49 - $150 +
A - 51 - $80 +

PAGE 36
A - 86 - $25 +
A - 92 - $50 +
A - 93 - $60 +
A - 138 - $30 +
A - 88 - $35 +
A - 59 - $30 +

PAGE 37
S6 - 17 - $75 +
S6 - 18 - $125 +

PAGE 38
#1110 - $50 +
#1111 - $50 +
#9051-10 - $75 +

PAGE 39
ALL $100 +

PAGE 40
#2602 - $50 +
#2607 - $125 +
#2601 - $30 +
#2626 - $150 +
#2606 - $125 +

PAGE 41
ALL $50 +

PAGE 42
ALL $50 +

PAGE 43
#2602 - $50 +
#2607 - $125 +
#2601 - $30 +
#2626 - $150 +
#2606 - $125 +

PRICE GUIDE

PAGE 44
#95 - $60 +
#57 - $40 +
#95 - $60 +
#50 - $40 +
#95 - $60 +
#118 - $60 +
#66 - $40 +
#150 - $100 +
#140 - $100 +
#153 - $40 +
#160 - $75 +

PAGE 45
#417 - $50 +
#U780 - $50 +
#224 - $50 +
#2518 - $100 +
#E654 - $100 +
#BRK2516 - $100 +
#BRJCB460 - $50 +
#2502 - $50 +
#AWL454 - $60 +

PAGE 46
#1411 - $75 +
#1400 - $25 +
#1414/2315 - $75 +
#1412 - $75 +
#1410 - $75 +
#1416 - $50 +
#1415/2315 - $100 +

PAGE 47
#2505 - $100 +
#2510 - $75 +
#2507 - $100 +
#2515/2278 - $80 +

PAGE 48
#1411 - $75 +
#1406 - 4150 +
#1410 - $75 +
#95 - $60 +
#95 - $60 +
#95 - $60 +

PAGE 49
#9078 - $25 +
#9044 - $100 +
#9079 - $35 +
#9031 - $50 +
#9046 - $100 +
#9045 - $100 +

PAGE 50
#1508 - $125 +
P
PAGE 51
#1412 - $75 +
#1415 - $100 +
#505 - $100 +

PAGE 52
ALL $50 +

PAGE 53
#2202 - $100 +
#2205 - $100 +
#2201 - $50 +
#2207 - $150 +
#2216 - $100 +
#2206 - $100 +

PAGE 54
#1561 - $35 +
#1565 - $200 +
#1562 - $50 +
#1568 - $200 +
#1566 - $150 +

PAGE 55
#1063 - $75 +
#2631 - $40 +
#1062 - $75 +
#2621 - $50 +
#2632 - $75 +
#2622 - $75 +

PAGE 56
#1031 - $75 +
#7060 - $400 +
#1032 - $100 +
#1060 - $400 +
#7057 - $250 +
#7063 - $200 +
#7054 - $500 +

PAGE 57
#5020 - $300 +
#5025 - $400 +
#5030 - $300 +

PAGE 58
#97643 - $300 +
#97644 - $100 +
#97645 - $300 +
#97638 - $300 +
#97646 - $300 +
#97639 - $100 +
#97647 - $300 +
#97634 - $100 +

PAGE 59
#97640 - $300 +
#97648 - #300 +
#97641 - $300 +
#97649 - $150 +
#96900 - $400 +
#97642 - $400 +
#97635 - $75 +
#97637 - $50 +
#97636 - $100 +

PAGE 60
#80464 - $200 +
#80463 - $200 +
#80465 - $200 +
#80466 - $200 +
#80468 - $200 +
#80462 - $400 +
#80460 - $200 +
#70528 - $200 +
#70530 - $300 +
#80467 - $200 +

PAGE 61
3B1626 - $50 +
3B1620 - $200 +
3B1628 - $50 +
3B1624 - $100 +
3B1634 - $50 +
3B1630 - $50 +
3B1622 - $125 +
LANTERN - $50

PAGE 62
3B1584 - $25 +
3B1580 - $40 +
3B1588 - $50 +
3B1586 - $40 +
3B1592 - $40 +
3B1590 - $40 +
3B1582 - $125 +
3B1594 - $50 +
3B1596 - $50 +

PAGE 63
3B1608 - $100 +
3B1602 - $150 +
3B1606 - $60 +
3B1604 - $50 +
3B1612 - $75 +
3B1610 - $75 +
3B1614 - $40 +
3B1600 - $175 +

PAGE 64
L-3023 - $125 +
ALL OTHERS $100 +

PAGE 65
#2100 - $75 +
#2101 - $75 +
#2102 - $75 +
#2103 - $75 +
#2104 - $75 +
#2113 - $100 +
#2112 - $75 +
#2114 - $100 +
#2109 - $75 +
#2110 - $75 +
#2107 - $75 +
#2106 - $75 +
#2108 - $75 +

PAGE 66
ALL $100 +

PAGE 67
#3230A - $100 +
#3544B - $100 +
#3297 - $50 +
#920M - $100 +
#3201 - $50 +

PAGE 68
ALL $100 +

PAGE 69
ALL $100 +

PAGE 70
GROUP 30 - $500 +
28H203 - $40 +
28H204 - $30 +
28H211 - $20 +
L2653 - $40 +

PRICE GUIDE

PAGE 71
GROUP 40 - $500 +
28H201 - $60 +
L2653 - $50 +
L354 - $30 +
L46 - $30 +
L2629 - $200 +

PAGE 72
ALL $100 +

PAGE 73
#1880 - $200 +
#1874 - $50 +
#1873 - $75 +
#1881 - $100 +
#1836 - $200 +
#1843 - $200 +
#1844 - $200 +
#1881 - $100 +
#1873 - $75 +
#1836 - $200 +
#1888 - $100 +

PAGE 74
REFERENCE ONLY

PAGE 75
#E1616B - $50 +
#E1616G - $50 +
OTHERS - $5,000 +

PAGE 76
ALL $100 +

PAGE 77
L2926 - $50 +
ALL OTHERS $100 +

PAGE 78
L2629 - $200 +
L354 - $30 +
L2653 - $50 +
L46 - $30 +
28H202 - $50 +

PAGE 79
#1245 - $500 +
S - 975 - $250 +
S - 976 - $150 +
S - 978 - $225 +

PAGE 80
#6265 - $2,000 +
MILLER - $400 +

PAGE 81
D3080 - $2,000 +
#6330 - $2,000 +

PAGE 82
ALL $200 +

PAGE 83
#8901 - $40 +
#8909 - $50 +
#8600 - $75 +
#8051 - $60 +
#8010 - $125 +
#8020 - $125 +

PAGE 84
#2153 - $35 +
#1612 - $75 +
#1707 - $100 +
#4004 - 4125 +
#2226 - $50 +
#2400 - $25 +
#3540 - $100 +
#3451 - $60 +
#4357 - $75 +
#4940 - $30 +
#1478 - $30 +
#1655 - $30 +

PAGE 85
#6712 - $60 +
#5277 - $50 +
ALL OTHERS $75 +

PAGE 86
#3249 - $125 +
#3230 - $100 +
#3245 - $150 +
#3247 - $175 +

PAGE 87
ALL $50 +

PAGE 88
ALL $75 - $100

PAGE 89
ALL $100 - $125

PAGE 90
#2000 - $50 +
#5000 - $100 +
#2005 - $50 +
#5005 - $100 +
#5010 - $75 +
#5015 - $100 +

PAGE 91
#5911 - $75 +
#5187 - $50 +
#5912 - $100 +
OTHERS - $100 +

PAGE 92
#2205 - $50 +
OTHERS - $75 +

PAGE 93
ALL $30 - $50

PAGE 94
#2050 - $75 +
OTHERS - $30 - $50

PAGE 95
#1523 - $75 +
#1294 - $100 +
#1533 - $100 +
#1084 - $150 +

PAGE 96
ALL $100 +

PAGE 97
ALL $100 - $125

PAGE 98
#4000 - $1,000 +
#4005 - $1,000 +
#4010 - $1,000 +
#4015 - $1,000 +
#4020 - $1,000 +
#4025 - $1,000+
#4030 - $1,000 +
#4035 - $1,000 +
#4040 - $2,000 +
#4045 - $2,000 +
#4050 - $2,000 +
#4055 - $2,000 +

PAGE 99
#4065 - $1,000 +
#4070 - $750 +
#4075 - $2,000 +
#4080 - $2,500 +

PAGE 100
ALL $75 +

PAGE 101
ALL $60 +

PAGE 102
#836 - $75 +
#1068 - $60 +
#936 - $50 +
#632 - $80 +
#732 - $75 +
#832 - $50 +
#932 - $50 +

PAGE 103
#2002 - $100 +
#2003 - $75 +
#2004 - $200 +
#2006 - $100 +
#2008 - $100 +
#19 - $300 +
#21 - $400 +
#203 - $200 +

PAGE 104
ALL $300 +

PAGE 105
ALL $100 +

PRICE GUIDE
COLOR SECTION

BACK COVER - HANDEL - $3,000 +

FRONT COVER - HANDEL - $3,000 +

PAGE A
ALL $3,000 +

PAGE B
GLOBES - $50 +
ALL OTHERS - $25 - $30

PAGE C
TOP ROW - ALL $100 +
MIDDLE ROW - ALL $200 +
BOTTOM ROW - ALL $100 +

PAGE D
#6930 - $2,000 +
#6896 - $1,500 +

PAGE E
#6884 - $2,000 +
#6893 - $750 +

PAGE F
$2,000 +

PAGE G
#6688 - $2,000 +

PAGE H
#7088 - $2,000 +
#7091 - $750 +
#7073 - $1,500 +
#6892 - $400 +
#6989 - $300 +